BIOLOGY INVESTIGATIONS

James H. Otto
Albert Towle
W. David Otto

Holt, Rinehart and Winston, Publishers
New York • London • Toronto • Sydney

James H. Otto
was a biology teacher and head of the Science Department
at George Washington High School, Indianapolis, Indiana.

Albert Towle
is a professor of biology and supervisor of biology student
teachers at California State University, San Francisco, California.

W. David Otto
is a teacher of biology at John Marshall High School,
Indianapolis, Indiana.

Editorial Development
William N. Moore, Roger R. Rogalin, M. Jean Young

Editorial Processing
Margaret M. Byrne, Shelley L. Feiler

Art and Production
Fred C. Pusterla, Robin M. Swenson, Annette Sessa, Beverly Silver,
Donna Bogdanowich, Dorina Virdo

Product Manager
Laura Zuckerman

Advisory Board
Rhenida Bennett, John W. Griffiths, Henry May, David J. Miller,
Jonathan Permar, George Salinger, John Taggart

Researcher
Pamela Floch

Consultant
John Matejowsky

Safety Consultant
Franklin D. Kizer, Executive Secretary, Council of State
Science Supervisors

Printed in the United States of America

ISBN 0-03-056033-0
1234 059 98765432

To the Student: How to Use this Book

This edition of **Biology Investigations** is organized in a way that allows you to work as a biologist. There is a section on "Working as a Biologist" on page 1 that describes some biological methods. The organization of each investigation supports these methods.

For each investigation, a *Problem* is presented. This is the same as the hypothesis in a scientific method.

The *Objectives* determine what should be learned from answering the questions presented in the problem.

A list of *Materials* tells you what equipment you will need to proceed with the *Investigations* section.

The *Analysis* section gives a background to the investigation. It is a thought-provoking introduction to an understanding of the problem presented. This section represents the research part of the scientific method. Along with information presented in the *Problem*, this section gives you a body of knowledge about the subject to be investigated. The data collected in this section is from textbook research and, in some cases, information based on the use of models. Models are used by scientists to represent an actual situation when the actual situation is impractical or inconvenient to use. For example, a tossing of coins in lab 10 represents the frequency of boys and girls in a population. Obviously, it would take several generations to answer the questions asked in the lab if actual individuals were used.

The *Investigation* section employs an inquiry approach that provides an opportunity to discover information in a laboratory. Models are also used in this section, but most investigations involve procedures that will give you specific data developed from the actual situation.

Recording of all data in **Biology Investigations** is done on the *Data Sheet*. You will follow the same procedure as will the other students, but since you are inquiring into the field of biology, you will encounter many variables. When you dissect a specimen, it will be similar to, and at the same time different from, another student's specimen. Your *specific* results are recorded on your *Data Sheet*. Data may be recorded in various ways: charts and graphs, diagrams and illustrations, drawings and questions to answer.

The *Summary* section contains questions that you should be able to answer after completing the Procedure and Data Sheet. The *Investigations On Your Own* section provides suggestions for further research.

The first lab in the book is called "Safety." Complete this lab carefully before doing any of the other investigations in the book. There are several occasions in the biology laboratory when caution is important. Where a potentially hazardous situation may exist, **Biology Investigations** has included various safety symbols that caution you to be on the alert.

These safety symbols are:

electrical safety

poison

gas precaution

explosion danger

hand safety

eye safety

animal safety

plant safety

fire safety

clothing protection

Look for them and, when you see them, follow the guidelines for safety presented in the "Safety" lab.

Contents

Working as a Biologist

As a biology student you will spend much of your time in the laboratory. It is here that you will discover the structures and functions of living things and their relationships to one another. You will learn to use some of the methods employed by scientists: experimentation, observation, collection and interpretation of data, and drawing of conclusions. Your techniques and ability to use laboratory apparatus should improve with each investigation you perform. At the same time, your knowledge of living things will increase.

Conducting a Biological Investigation

In each Investigation there is a problem to be solved. By fulfilling the stated objectives you should be able to answer the questions presented in the *Problem* section. The investigations are divided into parts. Your teacher will tell you whether you will be expected to do all of the parts or only certain ones. The length of time required to complete each part will vary. In addition to the assigned parts, you may also use the *Investigations On Your Own* for special projects or independent research.

Recording Biological Data

Biological data include all observations you make during an investigation. The data you record will be in several forms. In dealing with structure (the parts of living things), you may record your observations by labeling a drawing that has been provided. Sometimes you may be asked to draw your observation as well as label it. A label should always be printed and have a straight guideline running to the structure being named. The guidelines should be clear and should not cross each other, since this would cause confusion. Names of parts are always printed horizontally.

Most of the drawings you will be asked to make are representative drawings. These show the size, shape, and location of structures found in an organism. The drawings you make need not be artistic; few of us are artists. But drawings must always be neat and accurate or they have no scientific value. Data are often interpreted in graphs. Bar graphs may be used to illustrate relative quantities of various trials. Line graphs are used to show relationships between variables. For example, the growth of a yeast population can be illustrated with a line graph. (See Investigation 51.)

The subject of an investigation may deal with the behavior of an organism (how it reacts and what it does). Sometimes the investigation will deal with the function of an organ. Your observations will usually be in the form of answers to direct questions, data to record, or tables to complete. These answers should be based on facts gained from observation and research (including reading). Guesswork must never appear as scientific data. Although the results of your own experiment are not always what you expect, remember that the experiment that does not seem to work is often just as important to research as the one that does.

Preparing a Laboratory Report

Writing a lab report is very different from making observations and recording data in your laboratory manual. In doing a lab report, your teacher may give you an outline or you may develop one yourself. In general, however, it should include:

I. *Title* — This should be specific. "Growing Plants" or "Plant Nutrition" are too vague as titles. The title should tell exactly what you are studying. A good title might be: "The Effect of Mineral Deficiency on the Growth of Coleus Plants."

II. *Purpose* — This section should state the problem you are investigating. The statement should be simple. For example: How does the lack of certain minerals in the soil affect the growth of Coleus plants? Your entire report must relate to this problem.

III. *Materials and methods* — You must tell exactly what you did to prove or disprove the hypothesis. Be sure to have a control for your experiment. In this one, the control plant should be identical except for the mineral content of the soil. Describe the exact experimental procedures and use of minerals in the sequence in which you did them.

IV. *Observations and results* — This part forms the basis of support for your analysis and conclusions. It is purely objective. *Do not* include your interpretation as an observation. You must be sure that what you observe and measure is actually there. Do not overlook any result. Record *all* experimental data and observations you make. Give a complete description of what happened. Illustrations, drawings, graphs, sketches, and data tables should be used to support the information whenever possible.

V. *Analysis and conclusions* — These are subjective in nature. Here you can interpret your results. Tell how your results prove or disprove your original hypothesis. Make each conclusion positive and distinct. You should leave no doubt in the mind of the reader as to why you feel justified in making your conclusions on the basis of the evidence you have collected.

NAME DATE

Safety

You are about to embark on a series of laboratory investigations designed to give you "hands-on" knowledge of biology. In order to make your laboratory a place of learning and discovery, it must be a *safe place* in which to work. Safety in the lab is part of the scientific process of orderliness and logical thinking.

Carefully read the following guidelines as a preparation for your laboratory assignments. Then examine the drawing on the next page. Can you distinguish between correct, safe procedures and wrong, hazardous ones? Answer questions 1–15.

Safety Guidelines

(A) In General

... Approach each experiment with care and respect. No rough play of any kind can be tolerated.

... Never perform an experiment unless it is prepared and assigned by your teacher.

... Always note any possible cautions or hazards described in the laboratory material. Follow any safety instructions given by your teacher.

... Maintain sanitary conditions in the laboratory by cleaning equipment and work surfaces after use. Do not eat or drink from laboratory equipment.

... Tie back long hair. Avoid wearing loose, flowing clothing. Wear laboratory aprons.

... Take an interest in safety conditions in your lab. Help your teacher to make periodic checks of the laboratory and its equipment.

... Handle toxic or flammable substances with extreme care. Follow the advice of your teacher as to the use of these substances within a particular experiment. Do not taste any chemical substance. Keep combustible materials away from an open flame.

... Be sure all gas jets, electrical connections, and water faucets are turned off at the end of the laboratory period.

(B) Specifics

... Use artificial or indirect lighting for microscope observations. Direct sunlight can damage the eye.

... Use scalpels or razor blades that have only one cutting edge.

... Dissecting specimens should be properly mounted. Never hold a specimen while dissecting it.

... Kill bacterial cultures before washing Petri dishes by heating them for 20 minutes at 138 kPa of pressure.

... Sterilize contaminated culture media with a strong disinfectant and wash thoroughly.

... Flame wire loops before and after transferring microorganisms.

... Always use valved bulbs when transferring liquids with a pipette.

... If live animals are used, follow the care instructions for the animal. Handle the animals gently.

(C) First Aid and Emergencies

... Know the location of fire extinguishers, safety showers, fire blankets, and first aid kits.

... Report all accidents, even minor ones, to your instructor immediately.

... Establish safety procedures for various kinds of accidents, e.g., fire, chemical spills, etc.

SAFETY SHOWER
FIRE
SAFETY BLANKET
SMALLEY

NAME DATE

Safety

The drawing on the opposite page depicts a jumble of safe and unsafe procedures in the school laboratory. Answer the following questions based on this illustration and the safety guidelines information.

(1) List *3* unsafe activities shown in the laboratory drawing. ____________

____

____

(2) Explain why each activity is unsafe. ____________

____

____

(3) List *3* correct lab procedures depicted in the drawing. ____________

____

____

____

(4) Explain how each procedure would help to make the laboratory safe. ____________

____

____

(5) Name *2* safety guidelines that are not illustrated. ____________

____

____

Now read each statement and print *T* for true or *F* for false on the line.

____________ **(6)** The biology laboratory is a good setting for practical jokes.

____________ **(7)** Contaminated culture media needs to be sterilized with disinfectant.

____________ **(8)** Spilled chemicals should be wiped up immediately.

____________ **(9)** It's unnecessary to report minor laboratory accidents.

____________ **(10)** Performing experiments not assigned by the teacher is good creative procedure.

____________ **(11)** No chemicals should ever be tasted.

____________ **(12)** Pipettes for transferring liquids should be used with valve bulbs.

____________ **(13)** Only the teacher needs to know the location of the first aid kit.

_____________ **(14)** Artificial light should be used in microscopic observations.

_____________ **(15)** Scalpels or razor blades used in the lab should have only one cutting edge.

In the space provided, sketch your biology laboratory. Identify the location of the fire extinguisher, safety blanket, safety showers, and first aid kit. Also indicate any live animals in your lab, and itemize some of their care requirements. Label gas jets, electrical outlets, and water faucets or fountains. Note where chemicals are stored.

What procedure would you follow in each of these situations?

(16) Clothing on fire. ____

(17) Chemical spill on hand. ____

(18) Particle in your eye. ____

(19) Cut from broken glass. ____

(20) Animal bite. ____

NAME DATE

The Microscope 1

MATERIALS

ANALYSIS▶
- microscope, lens paper, cheesecloth

INVESTIGATION▶
- microscope
- lens paper
- cheesecloth
- slide
- cover glass
- medicine dropper
- clean cloth
- tap water
- textile fibers
- newsprint
- dissecting needle
- cotton thread (2 different colors)

OBJECTIVES▶
- To become familiar with the parts of a microscope
- To learn how to use the microscope

Problem

How is the microscope used?
The microscope is the basic tool in the study of biology. By means of lenses, the microscope magnifies objects and living things too small to be seen by the unaided eye.

Two principles involved in using the microscope are: magnification — the ability to enlarge an image, and resolution — the ability to produce a clear image. As magnification increases and light rays are bent more, the resolution or distinctness of the image decreases.

Procedure
Part I. Analysis

To know how to use the microscope, you must first be familiar with its parts. Place the microscope on your lab desk with the arm of the microscope facing you. The base should be resting at least 10 centimeters from the edge of the table.

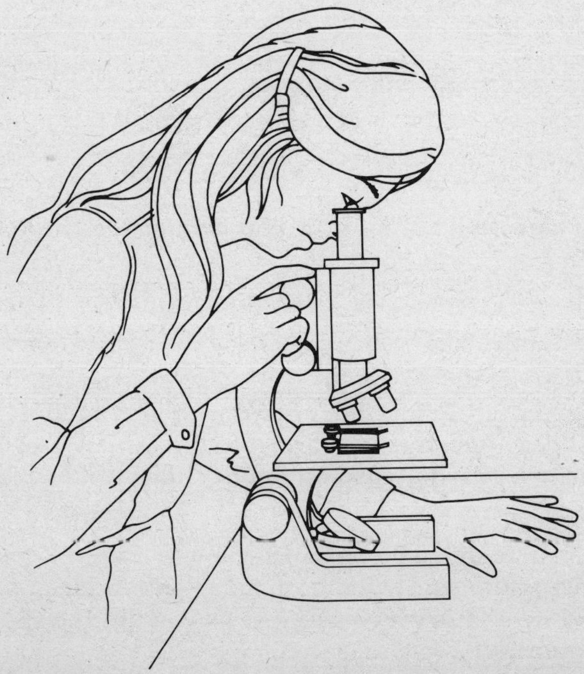

Label the drawing of the microscope on your Data Sheet as you examine its special features.

Use a piece of cheesecloth or other soft cloth to wipe off the *stage* and *frame* of the microscope. The stage is where the slide to be viewed is placed. *Do not use a cloth to wipe the lenses.*

Locate the *mirror* of the microscope. Wipe it with the cheesecloth. Turn the mirror so that the curved surface is facing a good light source. A microscope lamp, ceiling fixture, or daylight will provide sufficient light. Lighting is extremely important to a clear field of view. Some microscopes may have substage lighting. If this is so, the mirror is not needed. Never use direct sunlight as a source of light. This can be harmful to the eyes.

The eyepiece is the lens at the top of the microscope. The objective lenses are located on a revolving nosepiece at the bottom of the microscope tube. Use only special lens paper to clean the lenses. Answer questions 1–4 on your Data Sheet.

Locate the diaphragm of your microscope. The diaphragm regulates the amount of light passing through the specimen. There are two types of diaphragms — the iris and the disk types. Compare the diaphragm on your microscope with the drawings shown below.

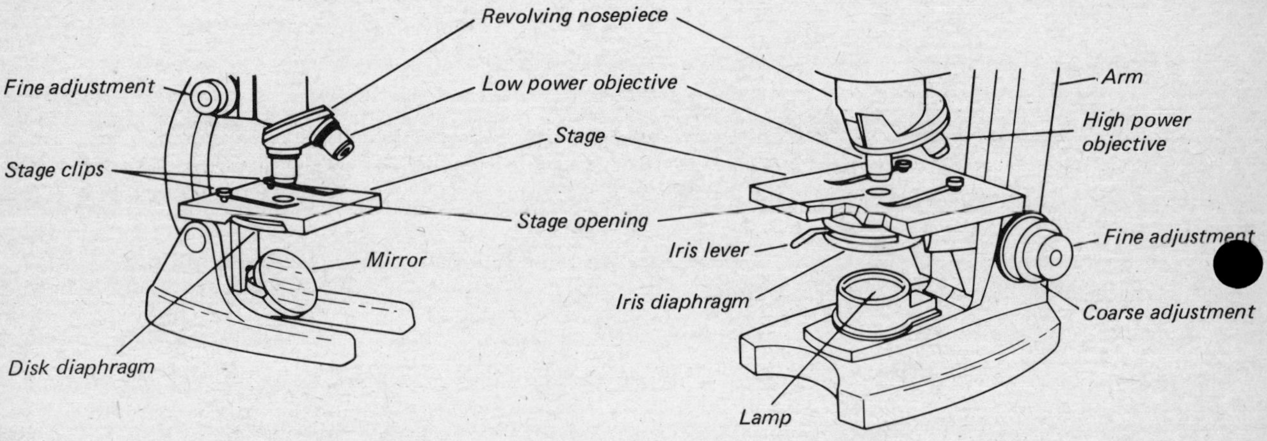

Open the diaphragm or turn the disk to the largest opening so that the greatest amount of light is admitted. Now you are ready to look through the eyepiece. Be sure to keep both eyes open.

Adjust the mirror and/or diaphragm if you cannot see through the eyepiece. Answer questions 5 and 6 on your Data Sheet.

Part II. Investigation

(A) Specimens observed through the microscope are mounted on glass slides. Wet mount slides are slides that you prepare for temporary use. Prepared slides are for permanent use. Your teacher may use prepared slides to illustrate certain subjects that you will study.

To prepare a wet mount slide of a newsprint lower case "e," follow these steps. Rinse a microscope slide with water and wipe both sides with a clean soft cloth. Hold the slide by the edges. Then rinse and dry a cover glass as you did the slide. Cut out a small piece of newsprint that contains a lower case "e."

Using a medicine dropper, place a drop of tap water in the center of the slide. Place the newsprint in the drop of water. Lower the cover glass as shown in the figure on the next page. This will prevent the formation of air bubbles. Place your temporary wet mount on the stage of the microscope and position it so that the letter "e" is facing you as you would read it. Clip it into place. Now you are ready to examine your "e."

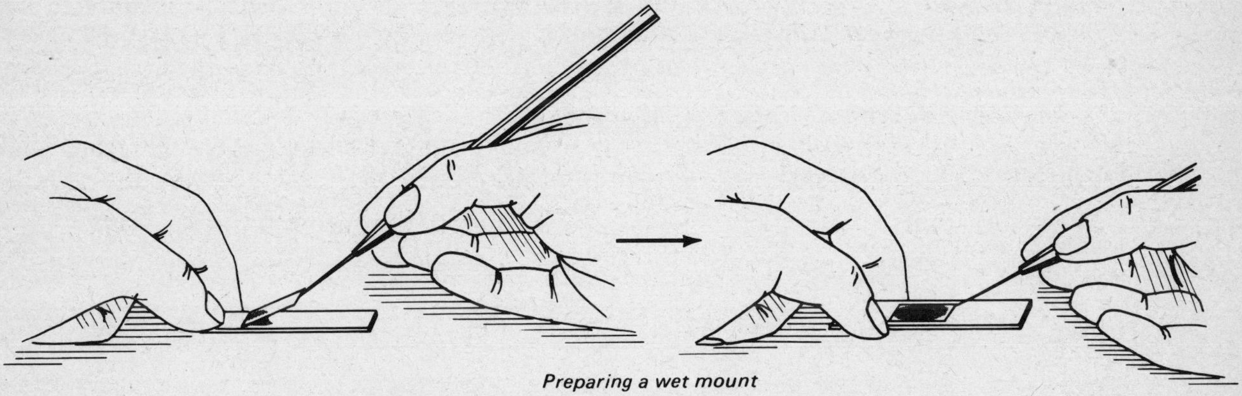

Preparing a wet mount

Clip your wet mount "e" specimen to the microscope stage. Using the *coarse adjustment*, lower the low-power objective as far as it will go without hitting the slide. Be sure to watch the bottom lens as you do this. *Never* lower the objective while looking through the eyepiece.

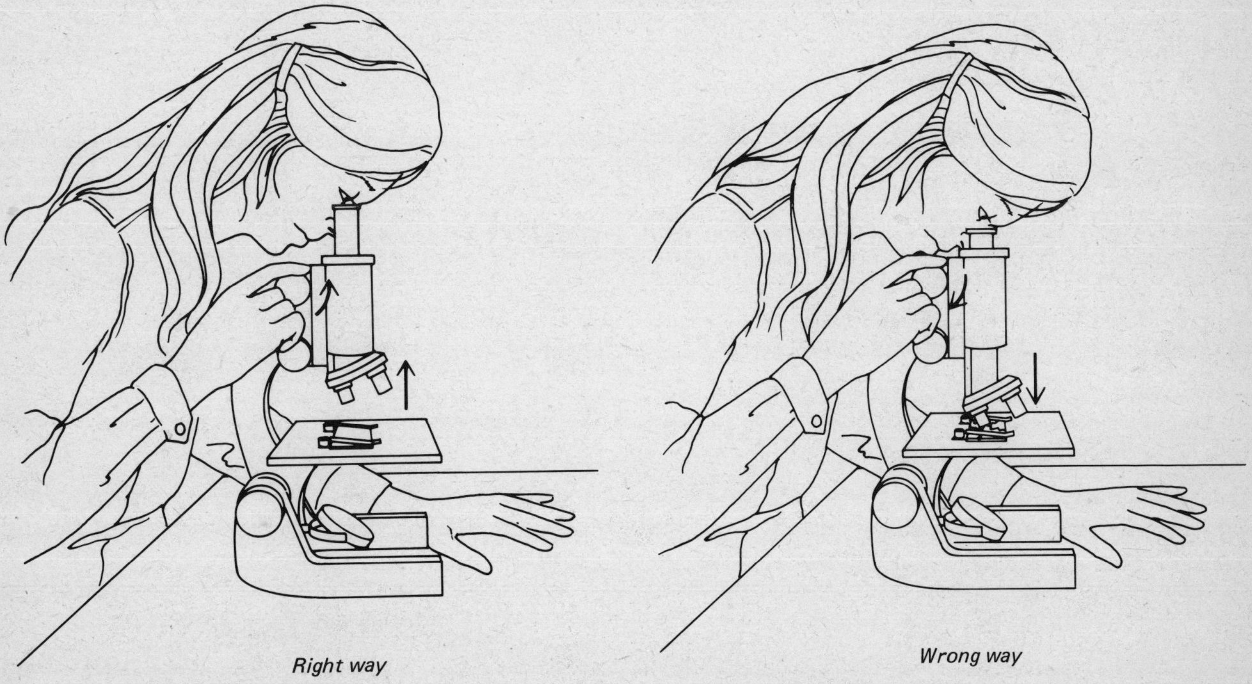

Right way *Wrong way*

Look through the eyepiece. Adjust the low-power objective by turning the coarse adjustment toward you. The letter "e" will soon come into view. Draw a picture of what you see under the microscope in the circle provided on the Data Sheet.

Move the "e" into the exact center of the low-power field. Turn the low-power objective out of the viewing position as you move the high-power objective into position. Turn the fine adjustment to achieve a sharp, clear focus. If you are having difficulty seeing the "e" clearly, check positions of the low-power objective and the slide. Draw a picture of what you see under the microscope in the circle provided on the Data Sheet. Answer questions 7–12.

(B) Because specimen materials have depth as well as length and width, you will need to shift your focus as you view in order to see details at various depths. Resolving power refers to how clear and sharp the magnified image can be made. Both light and resolving power are reduced as magnification is increased.

Mount 2 different-colored cotton fibers across each other in a drop of water on a microscope slide. Cover the slide with a cover glass. Be careful not to trap air under the cover glass. Adjust the diaphragm and bring a fiber into sharp focus with the low power of your microscope.

As you examine the fibers, shift the focus by turning the fine adjustment back and forth slowly. Answer questions 13–15 on the Data Sheet.

Move the fibers to the center of the low-power field and shift to the high-power magnification. Bring the fiber into the sharpest possible focus with the fine adjustment. Answer question 10 on the Data Sheet.

Investigations On Your Own

(1) Examine textile fibers such as linen, silk, wool, nylon, etc. under the microscope. How do these fibers differ? (Fiber identification is often important in criminal investigation.)

(2) Compare different types of paper — newsprint, tissue paper, stationery — under the microscope. What differences do you observe?

NAME DATE

The Microscope 1

DATA SHEET

Analysis

Label the parts of the microscope in the drawing below.

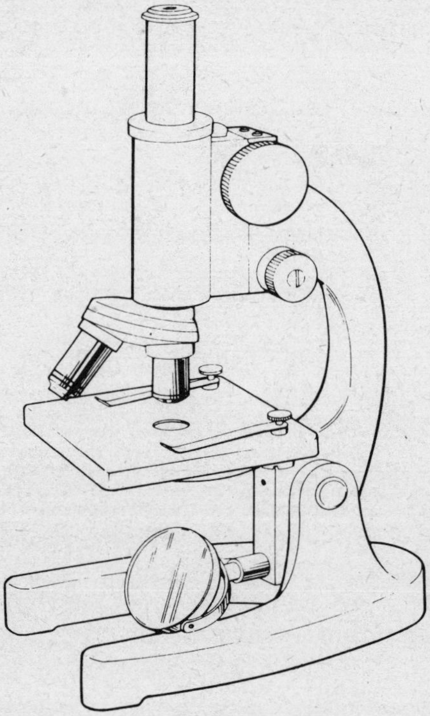

Standard compound microscope

(1) The low-power objective is shorter than the high-power objective. In what other way does the high-power objective differ from the low-power objective?

____

____

(2) When the low-power objective is directly over the opening of the *stage*, what magnification is printed on the lens? ______________________________

(3) What power is the eyepiece? ______________________________

(4) What is the total low-power magnification? ______________________________

(5) What type of diaphragm does your microscope have? ______________________________

(6) What do you see through the eyepiece? ______________________________

Investigation (A)

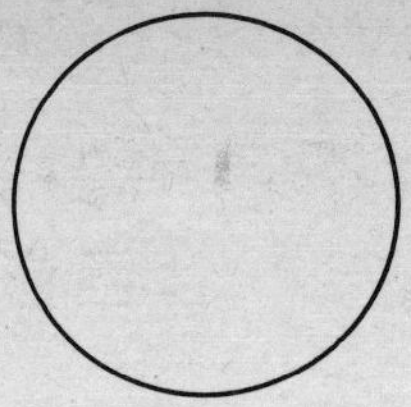

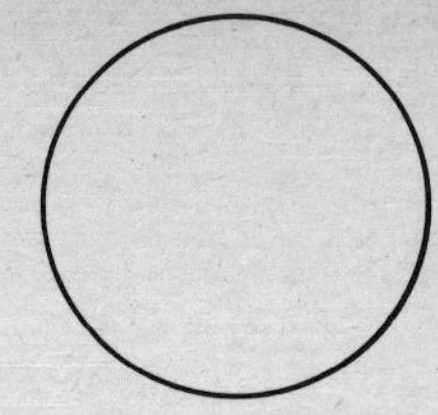

(7) Describe the position of the "e." ____________________

____

(8) What happens to the "e" as you move the slide to the right? ____________________

____

(9) What happens if you push the slide away from you? ____________________

____

(10) What is the total magnification of the high-power lens? ____________________

(11) About how many times was the magnification increased when you changed from low power to high power? ____________________

(12) How does this change the area of the slide included in the high-power field?

____

Investigation (B)

(13) Describe any changes in the appearance of the position of the fibers. ____________________

____

(14) Explain why these apparent changes occur. ____________________

____

(15) How can you determine which fiber is on top when you look through the microscope? ____________________

(16) What happens to the focus as you change from low-power to high-power objective?

____

Summary

Give the function of each of the following parts of the microscope:

eyepiece ____________________

stage ____________________

diaphragm ____________________

coarse adjustment ____________________

fine adjustment ____________________

low-power objective ____________________

high-power objective ____________________

mirror ____________________

NAME ____________________________ DATE ____________

Disproving Spontaneous Generation 2

MATERIALS

INVESTIGATION▶

- microscope
- slides
- cover glasses
- 6 flasks (250 mL)
- 2 solid stoppers
- 2 one-hole stoppers
- 8 cm section of straight glass tubing
- S-shaped glass tube
- 6 dropping pipettes
- pan for preparing broth
- autoclave or pressure cooker
- glass-marking pencil (heat resistant)
- aluminum foil
- nutrient broth (dehydrated) or beef broth
- distilled water

OBJECTIVES▶

- To appreciate the experiments of Spallanzani and Pasteur
- To become familiar with experimental procedures
- To be able to draw conclusions from scientific data

Problem

How was the theory of spontaneous generation disproved?

Spontaneous generation is the belief that living things can generate (spontaneously) from nonliving things. Because many great thinkers, including Aristotle, believed in spontaneous generation, it was accepted as a true theory. However, the experiments of Redi, Spallanzani and Pasteur proved that life comes only from life (biogenesis), and the controversial spontaneous generation theory was finally laid to rest.

Procedure

Part I. Analysis

Imagine yourself as the seventeenth-century scientist Francesco Redi. You are conducting a series of experiments to test the spontaneous generation theory. Compare your tests and results as depicted in the Data Sheet drawings. Then answer Data Sheet questions 1–7 to defend and support your findings.

Part II. Investigation

Set up six flasks as shown on the next page. Be careful to slowly turn the straight glass tubing and the S-shaped glass tubing through the one-hole rubber stoppers (flasks E and F).

Prepare 1000 mL (1 liter) of nutrient broth as directed on the media bottle. Be sure that all ingredients in the broth are dissolved. Mark the six flasks as A through F, using a heat-resistant glass-marking pencil. Divide the broth equally among the six flasks.

Prepare flasks A and B and set aside. Cover the mouths of C, D, E, and F with aluminum foil. Put these flasks and the stoppers for D, E, and F in the autoclave or pressure cooker. Sterilize at 15 pounds for 15 minutes.

After sterilization, allow the flasks and stoppers to cool in the sterilizer. Remove the aluminum foil from the flasks and insert the stoppers. Then place all the flasks in the incubator at 35–37°C for 72 hours.

Following the incubation period, examine all of the flasks. Using a clean pipette for each flask, put a drop of broth on a microscope slide. Add a cover glass and examine under high power.

If microorganisms are present, they will appear as small moving bodies. In some cases, it may be necessary to darken the field by slightly closing the diaphragm.

Answer questions 8 and 9, and fill in the table on your Data Sheet.

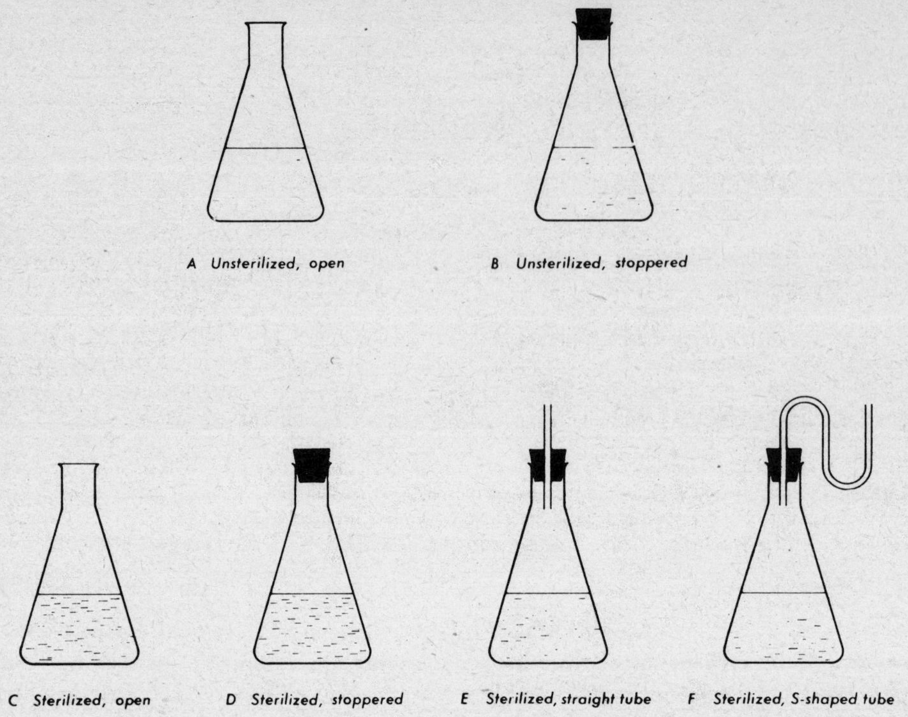

Method of preparing flasks A-F

Investigations On Your Own

(1) Construct an original investigation that disproves the theory of spontaneous generation. This investigation should be simple and should be done with easily available materials.

(2) Another theory related to the theory of spontaneous generation was plant growth through the conversion of soil into plant substances. One of the most famous experiments to disprove this theory was conducted by J. B. van Helmont in the seventeenth century. You may want to recreate this experiment by substituting a sunflower seedling for the willow tree. Be sure to draw your own conclusions after completing the experiment.

(3) Write a report detailing your findings and conclusions from your original investigation on spontaneous generation.

NAME DATE

Disproving Spontaneous Generation 2

DATA SHEET

Analysis

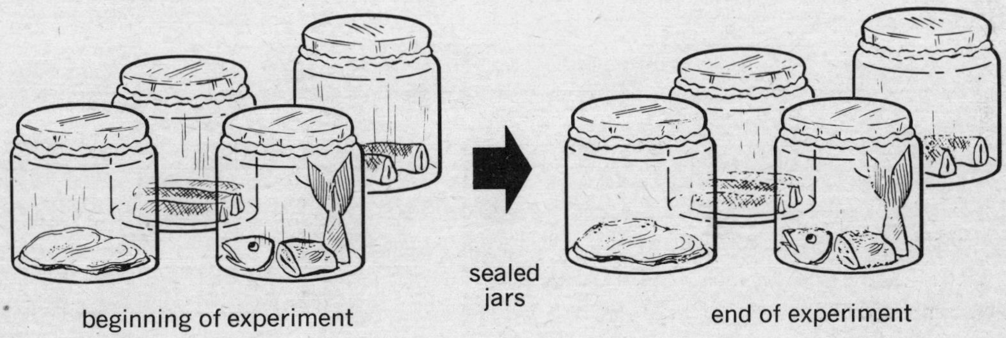

The experiments above were performed by Francesco Redi in 1668. Answer the following questions based on the drawings above. (Maggots and flies had often been observed on decaying food.)

(1) What problem did you try to solve by your experiments? ____________________

(2) Why did you choose decaying food for your tests? ____________________

(3) What was your hypothesis — your belief before your experiment?

(4) How did you decide to test your hypothesis? ____________________

(5) What happened to the opened jars? ____________________

(6) What happened in the sealed jars? ______________________________

____

(7) Based on the results of your experiments, what would you conclude?

____

____

Investigation

(8) Why is it necessary to sterilize the flasks and stoppers? ______________

____

(9) What is the purpose of incubating the flasks? ______________

____

In the table, fill in your observations for each of the six flasks. Each of the changes in the table indicates activity of microorganisms in the broth.

Flask	A	B	C	D	E	F
Presence of microorganisms						
Color						
Degree of turbidity (cloudiness)						
Odor						
Surface pellicle (film)						
Sediment						

Summary

Use the following key to summarize your findings from your investigation.

\+ evidence exists to prove — evidence to disprove

? possibility, but evidence is lacking 0 does not apply

Flasks	A	B	C	D	E	F
Living organisms were in the broth before incubation						
Broth substances were changed to living organisms by spontaneous generation						
Heat destroyed an "active principle" necessary for spontaneous generation						
Air is necessary for the growth of organisms						
Organisms present in the broth came from the air						
Organisms will not appear if neither the broth nor the air entering the flask contains organisms						

NAME DATE

Organic Compounds 3

MATERIALS

ANALYSIS▶
- tracing paper, scissors, construction paper

INVESTIGATION▶
- crucible
- tripod
- Bunsen burner
- test tubes
- test-tube holder
- test-tube rack
- white sugar
- glass-marking pen
- beaker
- brown wrapping paper
- Benedict's solution or Fehling's A and B
- Sudan III
- Biuret reagent
- iodine
- prepare solutions of:
  - corn syrup
  - 1% starch
  - vegetable oil or fat
  - 1% gelatin

OBJECTIVES▶
- To understand structural formulas and molecular models
- To identify the presence of carbon in organic compounds
- To use chemical tests to determine the presence of various organic molecules

Problem

What is the structure of organic compounds? How can the presence of organic compounds be detected?

All organic compounds contain the element carbon. Chemically, carbon is unique because of its ability to form 4 covalent bonds with other atoms, and its tendency to link with other carbon atoms in rings or long chains.

Living things depend on the energy from organic (carbon) compounds such as carbohydrates, proteins, and fats. Simple laboratory tests can identify the presence of specific organic compounds.

Procedure

Part I. Analysis

(A) Examine the structural formula of the sugar, glucose, on your Data Sheet. Note that glucose contains the elements carbon (C), oxygen (O), and hydrogen (H). Each letter represents one atom of the element. The lines between the atoms represent chemical bonds linking one atom to another. Answer questions 1–3 on your Data Sheet based on the information in the structural formula.

The glucose model (next to the structural formula) gives an idea of the molecular shape of glucose. Only the end atom of hydrogen and the end atoms of hydrogen-oxygen are shown in the model. Use tracing paper to cut out 3 glucose models; transfer the tracing to construction paper. Copy the dashed lines on the inside of the end atoms. Try joining two of the glucose molecules together. Now cut on the dashed lines and then try to join two glucose models together. (Save the end atoms.) Answer questions 4–7 on the Data Sheet.

(B) Sugars, such as glucose, are one kind of organic compound, called carbohydrate. Many sugar molecules joined together (in a chain) form another kind of carbohydrate called *starch*. Construct a part of a starch molecule by joining 3 of your glucose models together. Then answer questions 8–11 on the Data Sheet.

Part II. Investigation

(A) Both sugars and starches are carbohydrates. Have you ever noticed that when potatoes (starch) are burned, they turn black? This is a chemical property of carbon in the starch. In this investigation, you will determine whether carbon is also present in sugar crystals.

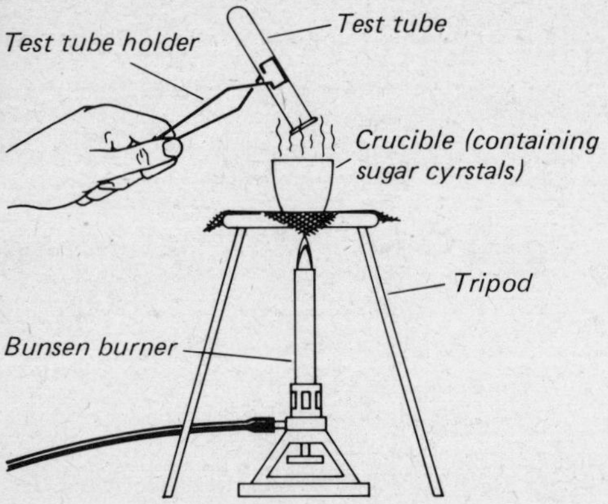

Place a small amount of sugar crystals in a crucible. Put the crucible on a tripod and heat with a Bunsen burner. Watch the sugar as you heat it.

Using a test-tube holder, hold an inverted test tube over the crucible. Let the tube cool in an upright position.

Continue to heat the crucible until you observe a black residue on the bottom. Answer Data Sheet questions 12–17.

(B) You can determine whether sugar, starch, fat, or protein molecules are present in a substance by using certain chemical tests.

Test for a simple sugar Using 2 marked test tubes, put 5 drops of water into tube *1* and 5 drops of corn syrup into tube *2*. Add 5 mL of Benedict's solution to each. Heat the tubes by placing them in a boiling water bath for at least five minutes. Observe any color changes. The final color in test tube *2* indicates the presence of simple sugar. This is a *positive* test. Answer questions 18–22.

Test for starch Using 2 marked test tubes, put 5 drops of water into tube *1* and 5 drops of starch solution into tube *2*. Add 2 drops of iodine to each tube. The color change in test tube *2* is a positive test for starch. Answer questions 23–25.

Test for fats and oils Spread 2 drops of olive oil on a piece of brown wrapping paper. Hold the paper up to the light. On another section of the brown paper, spread 2 drops of water. Answer questions 26–27.

Here is another test for fats and oils that is even more accurate. Put about one centimeter of the substance to be tested in a *dry* test tube and add 10 drops of Sudan III. Set the tube aside for 30 minutes. Report your finding by answering question 28.

Test for proteins Using 2 marked test tubes, put 20 drops of water into tube *1* and 20 drops of gelatin into tube *2*. Add 2 drops of Biuret reagent to each tube. (CAUTION: Use care in handling Biuret reagent. It may burn skin or clothing.) Observe the color changes and report them in your data sheet, questions 29–31.

Investigations On Your Own

(1) Test foods for the presence of carbon by heating them as you did the sugar crystals. Include the following foods in your experiment: cereals, fruits, maple syrup, candy, and cookies. Evaluate your findings. Do some foods seem to contain more carbon than others?

(2) Now that you know the positive test results for sugar, starch, fat, and protein molecules, test other common foods for these organic nutrients. Choose such food products as cheese, bread, potatoes, egg yolk, egg white, lunch meat, macaroni, apple, and soda. Keep a chart of your positive and negative tests for each sample. Try to explain any variations in your results.

NAME DATE

Organic Compounds 3

DATA SHEET

Analysis

(A) Examine the structural formula of glucose below and then answer the questions.

Structural formula of glucose

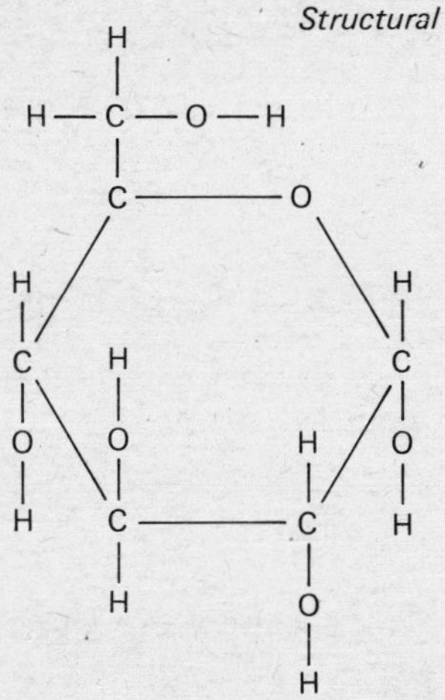

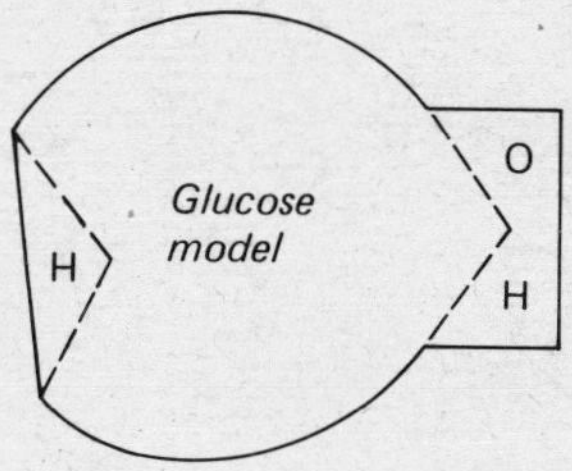

(1) What three elements are present in the sugar glucose? ______________________

____

(2) If each letter represents an atom of the element, how many atoms of carbon are in a molecule of glucose? ______________________

how many hydrogen atoms? ____________ oxygen atoms? ____________

(3) The molecular formula of water is H_2O — there are 2 atoms of hydrogen and 1 atom of oxygen in a molecule of water. Write the molecular formula of glucose by adding the correct subscripts. ______________________

(4) What happens when you try to join two glucose models (cut out on solid lines only)? ______________________

(5) What happens when you cut along the dotted lines and then try to join the glucose models together? ______________________

(6) What has been lost when joining the two glucose molecules together? ____________

____

(7) Two glucose molecules joined together form the double sugar maltose. Write the molecular formula for maltose. (Remember what is lost when two glucose molecules join.) ______________________

(B)

(8) Before you join the glucose molecules together, what must be removed from the model? ______

(9) Find the molecular formula for starch by counting the number of atoms in the middle glucose model. ______

(10) Compare the ratio of hydrogen to oxygen in sugar and starch.

______

(11) Dehydration means a loss of water. Why do scientists describe the joining together of sugar molecules as a dehydration process? ______

______

Investigation

(A)

(12) What do you observe in the test tube that was held over the crucible?

______

(13) What has happened to the hydrogen and oxygen atoms that were present in the sugar molecules? ______

(14) Describe what happens as you heat the sugar crystals. ______

______

(15) What do you think the black residue is? ______

(16) What type of change occurred in the sugar? ______

______

(17) Was there also an energy change involved? Explain your answer.

______

(B) Follow the directions for testing various organic nutrients as described on your procedure page. Record your results by answering the questions below.

Test for simple sugar

(18) What color appears in each tube? ______

(19) After heating, what color changes occur in tube *1*? ______

in tube *2*? ______

(20) What color finally appears in tube *2*? ______

(21) What was the purpose of tube *1*, which was a *negative* test? ______

______

(22) Describe a positive test for sugar. ______

NAME DATE

Test for starch

(23) What color change occurred in tube *1*? ____________________

(24) What color change occurred in tube *2*? ____________________

(25) Describe a positive test for starch. ____________________

____

Test for fats and oils

(26) What do you observe on the brown paper (with olive oil drops) as you hold it to the light? ____________________

(27) After allowing the water on the paper to dry, compare the water spot with the oil spot. ____________________

(28) Describe the test results using Sudan III. ____________________

____

Test for proteins

(29) What color change occurred in tube *1*? ____________________

(30) What color change occurred in tube *2*? ____________________

(31) Describe a positive test for protein. ____________________

____

Summary

(1) What is the one element common to all organic compounds? ____________________

(2) Why is the element you've just identified chemically unique? ____________________

____

____

(3) Name the three elements in carbohydrates. ____________________

(4) What is lost when sugar molecules join together to form double sugar molecules or starch molecules? ____________________

(5) The word carbohydrate is made up of carbon and hydrate (water). Why is the word carbohydrate a good description? ____________________

____

____

In the space below, draw the part of a starch molecule you constructed in Part I, Analysis (B). Indicate what was lost from the middle glucose model.

(6) What are the three elements in sugar? ______________________________

____

(7) How can you prove the presence of carbon in sugar? ______________________

____

Complete the following Table with your testing chemicals and results.

Organic Nutrient	Test Description	Test Result
sugar		
starch		
fats and oils		
proteins		

NAME DATE

4

Cells – The Basic Units of Life

MATERIALS

INVESTIGATION▶

- microscope
- slides
- cover glasses
- forceps
- bottle cork
- razor blade, single edge
- onion
- scalpel
- iodine stain
- *Elodea* leaves
- medicine dropper
- colored pencils
- human cheek cells
- toothpick
- methylene blue

OBJECTIVES▶

- To recognize the differences in structure between plant and animal cells
- To identify and observe cells – cork, onion, green plant and human epithelial cells

Problem

What are cells? What features distinguish plant cells from animal cells?

More than 300 years ago, the English scientist Robert Hooke first described the cavities in sliced cork as "cells." The cork cells Hooke examined were actually empty shells that had once contained live material.

In the early part of the 19th century, the German biologists Schleiden and Schwann realized that all living things — plants and animals — are made up of cells. This understanding eventually established the *cell theory*: the cell is the basic unit of structure and function of all living things.

Procedure

Part I. Analysis

Label the parts of the plant and animal cell drawings on your Data Sheet. Then complete your comparison by answering Analysis questions 1–3.

Part II. Investigation

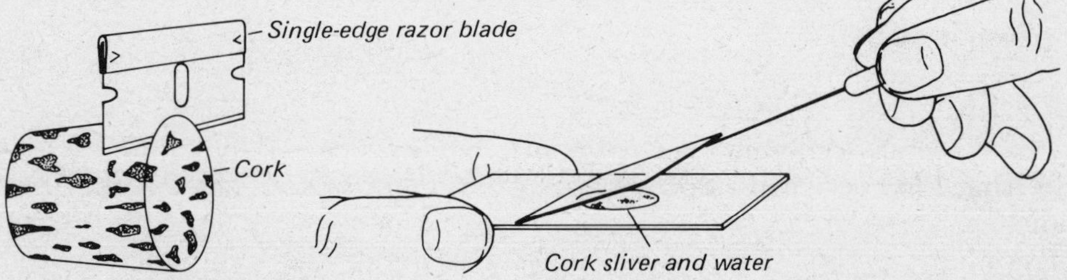

(A) Carefully shave a very thin section from a bottle cork with a single-edged razor blade. Prepare a wet mount of the cork slide. Examine the cork specimen under low power, studying it in different positions. Then examine the specimen under high power. (Remember not to use the coarse adjustment while viewing with the high-power objective.) Draw sketches of your observations in the spaces provided on your Data Sheet. Label the magnification powers by multiplying the eyepiece lens power by the low- or high-power lens number. Complete this section by answering questions 4–8.

(B) The epidermis of the onion is ideal for cell study because it is composed of a single layer of cells. As you study these cells, you are looking into functioning units of living material.

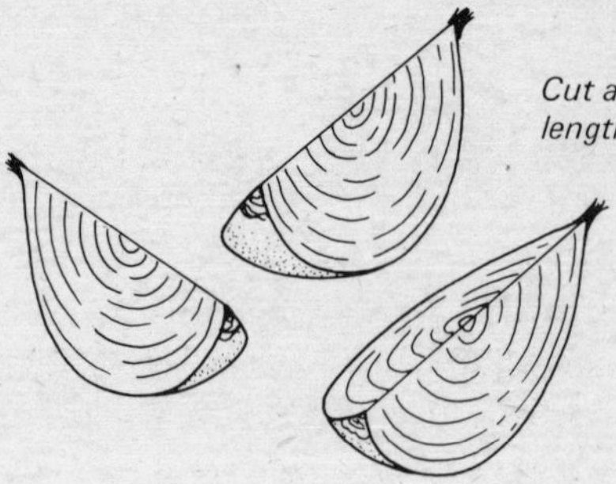

Cut an onion lengthwise

Cut an onion lengthwise. Remove a thick scale and peel the delicate transparent tissue from the *inner surface*. Cut a square of the tissue and mount it on a slide in a drop of water. (NOTE: Avoid wrinkling the tissue.) Add a cover glass to your wet mount. Then examine the living cells under low power. Answer questions 9–11.

Carefully raise one side of the cover glass and add a drop of iodine stain. Select one cell that shows the contents clearly. Move it to the center of the microscope field. Using high power, examine all the parts of the cell. Answer questions 12–16 and draw the onion cells under high power.

Peel transparent tissue from inner surface

(C) Prepare a wet mount of an *Elodea* leaf. The whole leaf should be used. Examine the leaf under the low power of the microscope. Then select a portion of the leaf where the cells are particularly distinct. Center this portion in the microscope field. Bring it into focus under high power. Use the fine adjustment to observe the cells at various depths.

Observe the small, oval, green bodies that appear in the cells. These are the chloroplasts. As you observe the chloroplasts, watch carefully for chloroplast movement. It may require several minutes of observation. Now answer questions 17–20 on your Data Sheet and draw some *Elodea* leaf cells in the space provided. If you have observed chloroplast movement, indicate the direction. Label cell wall, chloroplasts, cytoplasm, and nucleus in your sketch.

(D) In this section, you will examine the cell structure of human epithelial (cheek) cells. Note the differences between these animal cells and the green plant cells studied in section C.

Gently scrape the inside of your cheek with a clean toothpick. Prepare a wet mount of the material that you have scraped. Add a drop of methylene blue and a cover glass. Examine the cells under low power of the microscope. Switch to high power. Carefully look for the outer edge of the cytoplasm. Then answer questions 21–24. Draw a single cheek cell as you observed it under high power and label the plasma membrane, cytoplasm, and nucleus.

Investigations On Your Own

(1) Observe many different types of nonliving and living cells. Compare your findings to those you discovered in this Investigation. Draw sketches of the cells and their organelles.

(2) Examine the mitochondria of some cells under the light microscope. Place a cut celery stalk (containing "strings"), with the inner surface up, in a 5% sucrose solution. Cut a thin strip from the "strings." Then observe the mitochondria.

NAME DATE

Cells — The Basic Units of Life 4

DATA SHEET

Analysis

The outline drawing below is of a typical plant cell. Fill in the labels for the following plant cell features: nucleus, cytoplasm, cell wall, chloroplast, plasma membrane, and nucleolus. Find the location of the following plant cell features: *mitochondrion, vacuole, Golgi apparatus, endoplasmic reticulum, ribosome, lysosome, nucleolus, nucleus, cell wall, chloroplast, cytoplasm,* and *plasma membrane*. Add these labeled parts to your drawing.

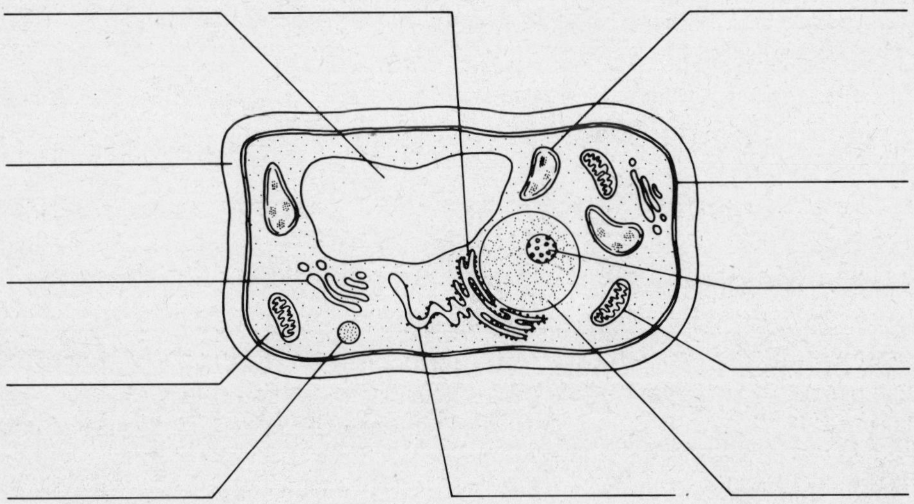

Now label the following parts in this typical animal cell drawing: nucleus, cytoplasm, plasma membrane, and nucleolus. Find the location of the following animal cell parts: *Golgi apparatus, lysosome, mitochondrion, endoplasmic reticulum, ribosome, nucleus, nucleolus, cytoplasm,* and *plasma membrane*. Add these parts to your drawing below.

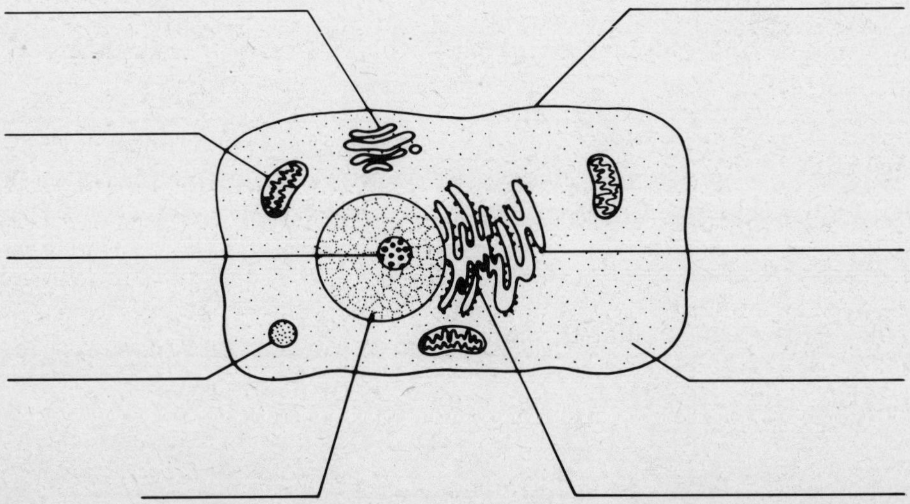

(1) Name the cell features found only in plant cells.

(2) Describe the functions of these parts.

(3) What advantages do plant cells have over animal cells?

Investigation

(A) Draw sketches of your cork specimen as you observed it under low power and under high power. Label the magnification in each drawing.

(4) How would you describe the units that compose the cork?

(5) Describe the shapes and sizes of the individual units.

(6) What do you see inside the units?

(7) Are there spaces between the cells?

(8) Are these cells living or nonliving? Explain.

(B) Examine the onion cells under the low power objective and answer the following questions.

(9) What is the shape of the cells?

(10) Are all the cells similar in shape?

(11) What color is the living cytoplasm?

NAME DATE

(12) When you add a drop of iodine, what effect does the stain have on the cells?

____

Choose one cell that is clear and switch to the high-power objective. Examine its parts and answer these questions.

(13) What is the appearance of the cytoplasm? ______________________________

(14) What is the appearance of the nuclei? ______________________________

(15) Are the nuclei always in the same position in the cell? ______________________________

(16) Does the onion epidermal cell have depth? Explain. ______________________________

____

Draw the onion cells under high power. Label the parts and magnification power.

(C) Observe an *Elodea* leaf with the high-power objective of your microscope. Then answer the following questions.

(17) In which layer are the widest cells located? ______________________________

(18) Note the oval, green bodies in the cells. These are the chloroplasts. Are any of the chloroplasts moving? ______________________________

(19) If you see movement, are all the chloroplasts moving in the same direction? at the same speed? ______________________________

(20) Explain how chloroplasts move. ______________________________

____

Draw some cells of an *Elodea* leaf in the space provided. Use arrows to indicate the direction of the chloroplast movement. Label your drawing, indicating the cell wall, chloroplasts, cytoplasm, and nucleus.

Draw some cells of an *Elodea* leaf in the space provided. Use arrows to indicate the direction of the chloroplast movement. Label your drawing, indicating the cell wall, chloroplasts, cytoplasm, and nucleus.

(D) Follow the procedure directions for scraping the inside of your cheek and preparing a wet mount slide. Examine the cells carefully with the low- and high-power objectives. Report your findings in the questions below.

(21) How does the outer edge of the cheek cells compare with the outer edge of the *Elodea* cells? ______________________________

(22) What is this outer edge called? ______________________________

(23) Describe the shape of cheek cells. ______________________________

(24) In what ways do the cheek cells differ from the *Elodea* cells?

____

____

Draw a single cheek cell as observed under high power. Label the plasma membrane, cytoplasm, and nucleus.

Summary

(1) How did the cork units differ from those of the onion, *Elodea*, and cheek cells?

____

(2) What are the differences between plant and animal cells? ______________________________

____

____

(3) What is the outer covering of a cheek (animal) cell called? ______________________________

(4) What is the basic unit of structure in all living things? ______________________________

NAME DATE

Selectively Permeable Membranes and Osmosis 5

MATERIALS

INVESTIGATION▶
- cellophane dialysis tubing
- glass rod
- string
- starch suspension
- 80% glucose solution
- distilled water
- 3 beakers
- medicine droppers
- 2 test tubes
- Benedict's solution
- Bunsen burner
- tripod
- iodine
- sprig of *Elodea*
- microscope
- glass slide
- cover glass
- 15% NaCl solution
- filter paper

OBJECTIVES▶
- To explain turgor pressure in plant tissues
- To observe how a selectively permeable membrane works
- To observe osmosis in cells placed in varying solutions of salt concentrations

Problem

What is a selectively permeable membrane? When does osmosis occur in living cells? Cells need to take in food, oxygen, water, and other vital molecules. They also need to excrete waste materials. Some substances are allowed to pass in and out — the plasma membrane surrounding the cell is *permeable* to those substances. But the plasma membrane is actually *selectively* permeable. Only certain substances are allowed to pass in and out.

Osmosis is the diffusion of water through a selectively permeable membrane. This occurs when there is an area of greater concentration and an area of lower concentration (of water). The water moves in the direction from greater to lesser concentration.

Procedure

Part I. Analysis

Turgor is the rigidity or stiffness of plants due to the water present in their cells. Study the drawings on your Data Sheet that depict the presence and loss of turgor pressure. Compare what happens to the vegetables soaked in distilled water with those soaked in salt solution. Also observe the vegetables that were not placed in either liquid. The drawings show the results of an actual experiment. Relate the conclusions by answering questions 1–7 on your Data Sheet.

Part II. Investigation

(A) In order to understand how a selectively permeable membrane works, it's helpful to make a model of a cell. Soak a 20-cm section of cellophane dialysis tubing in water for a few minutes. Rub the ends between your thumb and index finger until the ends separate. Insert a glass rod to hold it open. Twist one end and tie it tightly with string.

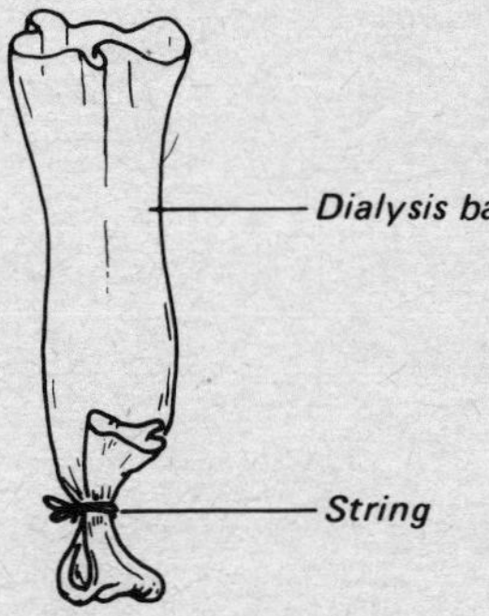

Remove the glass rod and then fill the dialysis bag ¾ full with starch suspension and ¼ full with 80% glucose solution. Tie the top of the bag with string. Leave a loose piece of string 10–15 cm long. Rinse the bag with distilled water. Place the bag in a beaker of distilled water. Leave the string outside the beaker so that you may remove the bag from the water. After 20 minutes, withdraw a dropperful of water from the beaker. Place this liquid in a test tube and add 10 drops of Benedict's solution. Heat in a boiling water bath for 10–15 minutes.

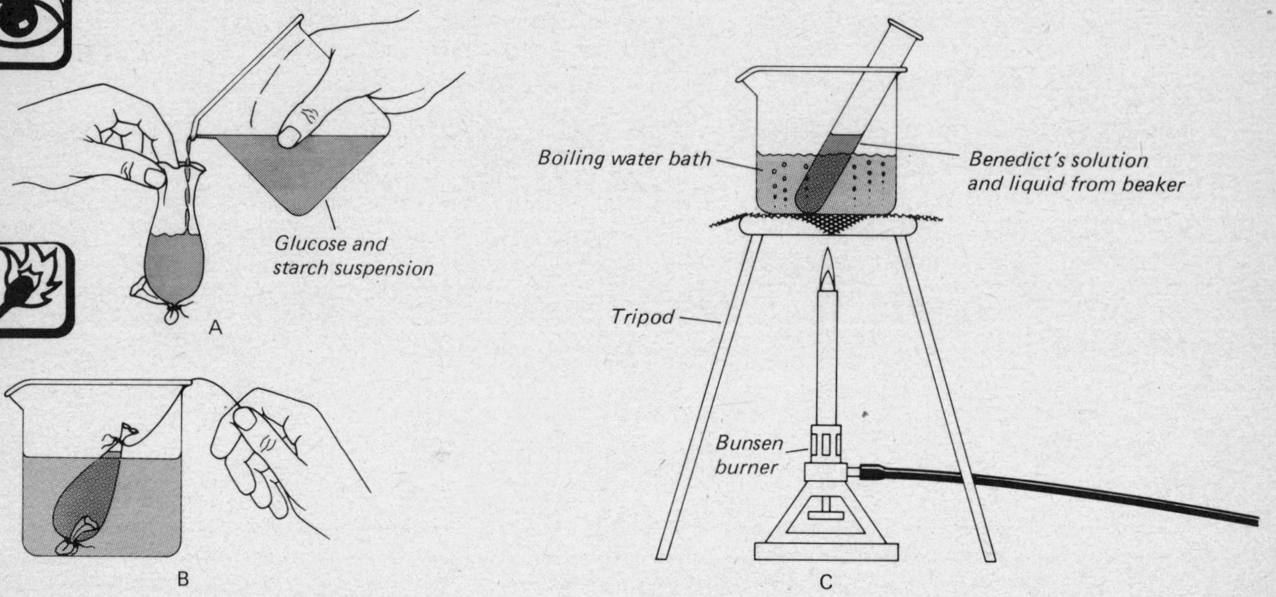

While the water bath is boiling, place 30 drops of the liquid from the beaker into a test tube. Add 3 drops of iodine. Observe the color of the water after the iodine has been added. Now report your results by answering questions 8–17.

(B) The movement of water into and out of cells can also be observed with the microscope. In this section, you will observe plant cells in solutions of varying salt concentrations.

Remove one leaf from a sprig of *Elodea* and prepare a wet mount slide with tap water. Examine it under high power. (Remember to center and focus it under low power first.) Draw the cell in the space provided on your Data Sheet page and answer questions 18–19.

Place one drop of 15% NaCl solution at the edge of the cover glass. About every 30 seconds, add another drop of NaCl solution until you observe a change in the cell. Draw the changed cell and answer questions 20–23. Now place a piece of filter paper at the edge of the cover glass and absorb as much of the salt solution as possible. Add distilled water at the edge of the cover glass, a drop at a time; watch the cells closely. Draw the cell and answer question 24.

Investigations On Your Own

(1) Study the action of the enzyme diastase on diffusion. Prepare a dialysis bag using diastase solution: add starch solution to the diastase and tie off the dialysis bag. Fill a beaker with distilled water (37° C). After 20 minutes, take two samples of the water from the beaker. Test one for starch and one for glucose. Record your observations and draw conclusions from your results.

(2) The permeability of cell membranes can be tested by using red blood cells. Prepare four test tubes of tap water, one for each of the following temperatures: 0° C, 25° C, 50° C, and 65° C. Mix three drops of blood in each test tube. Place a stopper in each tube and invert a few times. The bursting of red blood cells is called hemolysis. (You will know that this has occurred in your test tubes when the cloudiness becomes clear.) Write a report on your findings.

NAME ______________________ DATE ______________________

Selectively Permeable Membranes and Osmosis 5

DATA SHEET

Analysis

Study these drawings, which show the presence and loss of turgor pressure. Then answer the questions based on your understanding of the results.

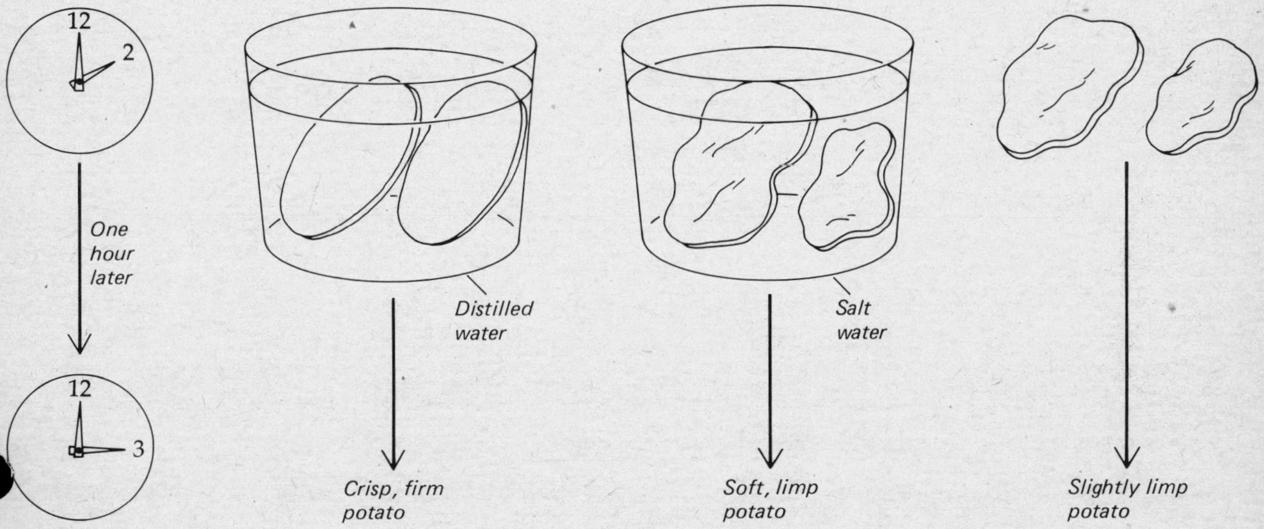

(1) How would you describe the stiffness and texture of the sliced vegetables (not placed in liquid) after an hour? ______________________

____

(2) How would you describe the stiffness and texture of the vegetable slices left in distilled water? ______________________

(3) What happened to the turgidity of these cells? ______________________

(4) How would you describe the stiffness and texture of the vegetable slices soaked in salt water? ______________________

(5) Do these slices have greater or lesser turgidity than the slices left out in the air?

____

(6) What conclusions can you draw from the results? ______________________

____

____

(7) Explain the movement of water in this experiment.

______

______

______

______

______

______

Investigation

(A)

(8) Why is it necessary to rinse the dialysis bag with distilled water? ______

______

(9) What is the purpose of heating Benedict's solution and the (beaker) liquid?

______

______

(10) What happens when you heat Benedict's solution and some water from the beaker?

______

(11) Explain your results. ______

______

(12) Do you note any changes in the amount of liquid in the dialysis bag? ______

______

(13) Is there a change in the amount of liquid in the beaker? ______

(14) What conclusion can you draw from this observation? ______

______

(15) What is the purpose of adding iodine to the liquid in the beaker? ______

______

(16) What happens when you add iodine to the liquid in the beaker? ______

______

______

(17) If a change has occurred, what does this indicate? ______

______

(B) Examine the *Elodea* leaf specimen with the high-power objective of your microscope. Report your findings below.

(18) What cell structures do you see in your *Elodea* specimen?

(19) Describe the position of the cell membrane and chloroplasts.

In the space provided, draw and label the parts of the *Elodea* cell that you observed.

(20) How many drops of NaCl were added before a change was noticed?

(21) What happened to the water in the cell?

(22) What visible proof do you have for your answer to (14)?

(23) Why do you think this change occurred?

In the space provided, draw the cell that you observed after the NaCl was added.

(24) After you absorbed some of the salt solution and added distilled water, do you observe any changes in the cells? ______________________________

____

In the space provided, draw the cell(s) you observed after the distilled water was added.

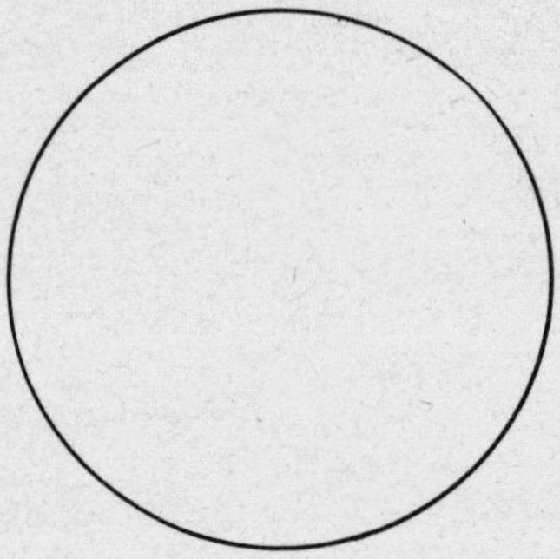

Summary

(1) How would you explain the condition of the vegetable slices that were in the salt solution? ______________________________

____

(2) How does a selectively permeable membrane work? ______________________________

____

(3) Explain the difference between osmosis and diffusion. ______________________________

____

____

NAME DATE

Photosynthesis and Cellular Respiration 6

MATERIALS

ANALYSIS▶
- tracing paper, scissors, construction paper.

INVESTIGATION▶
- *Anacharis (Elodea)*
- 7 test tubes
- 4 pieces of glass rod (5 cm)
- test tube racks
- 150-watt reflector lamp
- sodium bicarbonate
- bromothymol blue solution
- aquarium water
- graduated cylinder

OBJECTIVES▶
- To compare the reactions of photosynthesis and cellular respiration
- To observe the rate of photosynthesis under different light conditions
- To demonstrate that carbon dioxide is produced during respiration

Problem

How does light affect photosynthesis? What is produced during cellular respiration? Photosynthesis is the process of making food (usually in the form of glucose) from carbon dioxide and water in the presence of light. This complex process of changing light energy into chemical energy takes place in green plants — plants that contain chlorophyll. The rate of photosynthesis is determined by the amount of light available. As light intensity increases, the rate of photosynthesis increases.

Cellular respiration is the main energy-producing process in living cells. It involves the breakdown of food (glucose) and the oxidation of food molecules. Carbon dioxide and water are given off in cellular respiration. But the most important product of respiration is energy in the form of the ATP molecule. All living things — animals *and* plants — depend on the energy of cellular respiration.

Procedure

Part I. Analysis

(A) ATP, or adenosine triphosphate, the chemical energy produced in cellular respiration, is made of the molecules adenine, ribose, and three phosphates. (Adenosine refers to adenine and ribose, triphosphate to the three phosphates.)

Examine the structural formulas and models for each part of the ATP molecule on your Data Sheet page. Trace and cut out construction paper models of adenine (1), ribose (1), and phosphate (3). Attach the five models to form adenosine triphosphate. What needs to be removed in order to attach the models?

When energy is needed by an organism, the last phosphate group is separated from the ATP molecule and adenosine diphosphate (ADP) is formed. Energy is given off from the high-energy bond between the second and third phosphate groups. Make the ADP molecule by detaching the last phosphate group.

ATP can reform by the addition of phosphate to ADP. The energy needed (to change ADP to ATP) comes from the breakdown and oxidation of glucose. Food and oxygen are absolutely essential to life. Use your ATP-ADP models to answer questions 1–7.

(B) Trace and cut out models for glucose, water, oxygen, and carbon dioxide. Combine these models with the ATP model from **(A)** to depict the simplified equation for photosynthesis. Then rearrange the molecules to depict the equation for cellular respiration. Also indicate the number of each molecule in a balanced equation. Answer Data Sheet questions 8–12.

Part II. Investigation

(A) The rate at which photosynthesis takes place can be determined by recording the number of oxygen bubbles that escape from the cut end of a sprig of *Anacharis*.

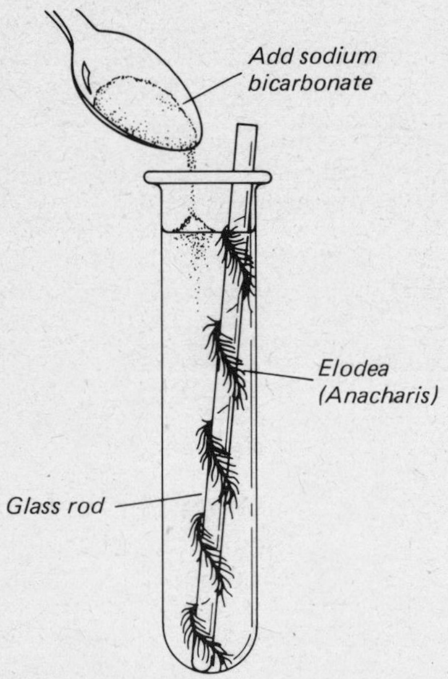

Fill 4 test tubes with water to within three cm of the top. Select 4 sprigs of *Anacharis* that have been growing in bright light. Each sprig should be about 7 cm long. Wind each sprig around a glass rod and immerse each in a test tube. Add a small amount of sodium bicarbonate to each test tube.

Place one test tube under bright light or a reflector lamp, the second test tube in daylight, the third in diffuse light, and the fourth in total darkness. After about 15 minutes, observe the cut ends of each plant. Count the number of oxygen bubbles given off by each plant in one minute. Then count the number of bubbles in five minutes. Record your results by filling in the Data Sheet Table and answering questions 13–18.

(B) It is possible to demonstrate that plants respire by testing for the presence of carbon dioxide. Bromothymol blue is blue in an alkaline environment, but it turns yellow in an acid environment. (Carbon dioxide in the presence of water forms a weak acid). Prepare a solution containing 50 mL of aquarium water and 20 mL of bromothymol solution. Fill 3 test tubes to within 3 cm of the top with this solution. Add a sprig of *Anacharis* to 2 of the test tubes.

Place one test tube with a plant and one test tube without a plant in the dark. Put the other test tube in bright sunlight. Observe the tubes the next day. Answer questions 19–22 on your Data Sheet.

Investigations On Your Own

(1) Determine the effect of temperature on the rate of photosynthesis. Prepare 4 sprigs of *Anacharis* as you did for Part II, **(A)**. Place one test tube in a water bath and use ice chips to reduce the temperature to 10° C. Record the oxygen bubbles given off after 1 minute and after 5 minutes. Add hot water to the water bath and separately raise the temperature to 20° C, 30° C, and 40° C. Use a separate test tube at each temperature. Record the number of oxygen bubbles released. What are your conclusions about the effect of temperature on photosynthesis?

(2) Find out whether germinating seeds release carbon dioxide. Soak some seeds (wheat, corn, or peas) in 5% formalin solution for about 20 minutes. Wash the seeds in cold water and soak overnight. Fill a gas bottle one-third full with seeds. Fill a second bottle about one-third full with bromothymol blue. Use a U-tube to connect the two gas bottles. Set up a similar apparatus as a control. However, destroy the seeds in the control by soaking them in 10% formalin for an hour. Observe what occurs in the demonstration bottle and in the control. Draw conclusions from your observations.

NAME DATE

Photosynthesis and Cellular Respiration 6

DATA SHEET

Analysis

(A) Look at the structural formulas and models for the parts of the ATP molecule. Assemble a model of ATP (and ADP) by following the directions on your Procedure page. Then answer the questions.

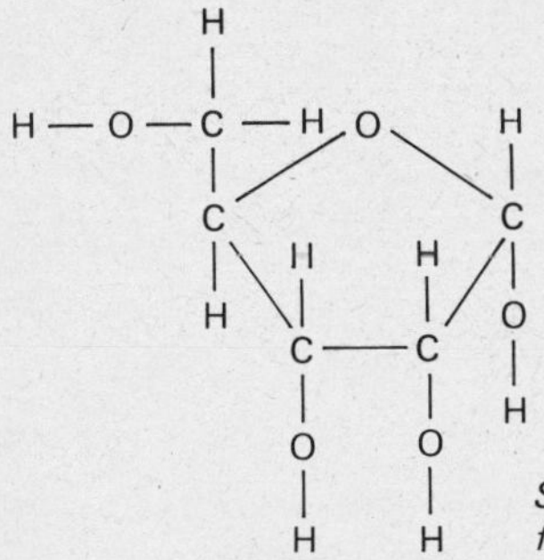

Structural formula for ribose (sugar)

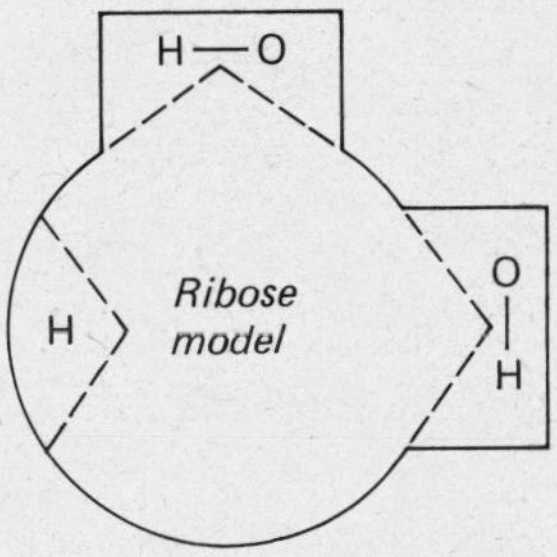

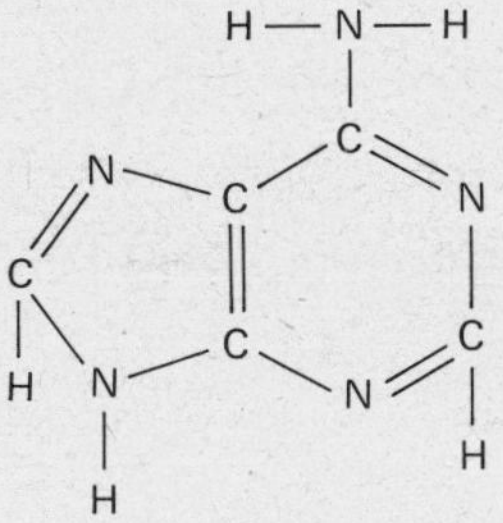

Structural formula for adenine

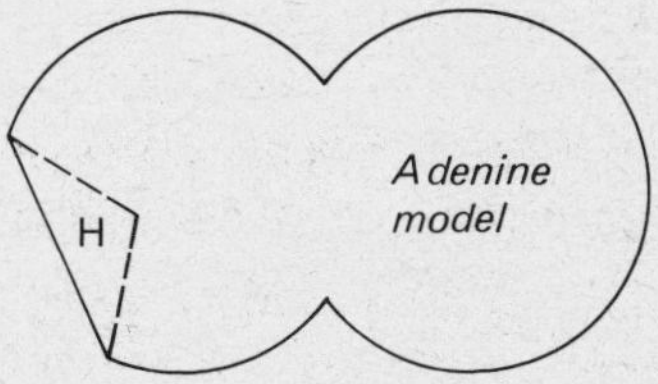

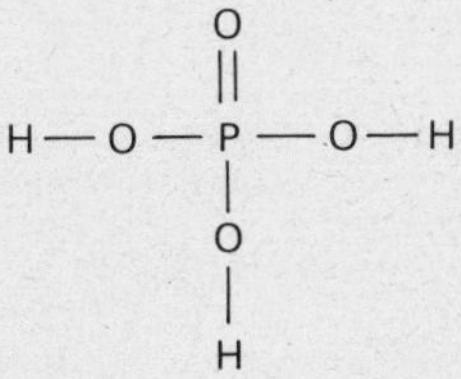

Structural formula for phosphate

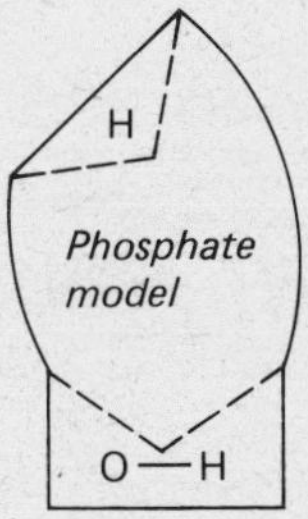

(1) Name the parts of the ATP molecule. ______________________________

____

(2) What must be removed when you link adenine with ribose? ______________

(3) What is formed when adenine and ribose are joined? ______________

(4) What does the prefix "tri" refer to in ATP? ______________

(5) When you remove the last phosphate from ATP, what is formed? ______________

____

(6) What is given off when the last phosphate is separated from ATP? ______________

____

____

(7) A constant supply of energy is needed by living things. What energy helps to reform ATP from ADP and phosphate? ______________________________

____

(B) Trace and cut out the models below to arrange them for photosynthesis and cellular respiration. Read the instructions on your Procedure page.

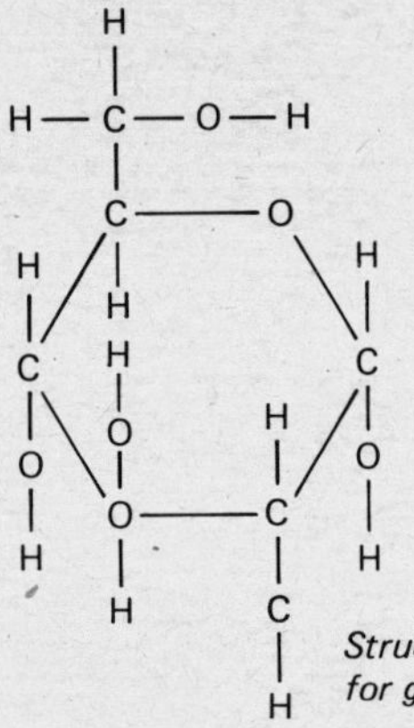

Structural formula for glucose

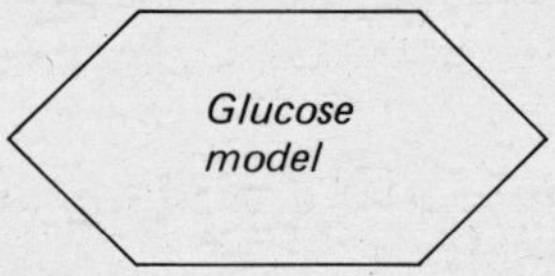

H—O—H

Structural formula for water

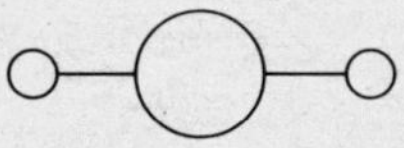

Water model

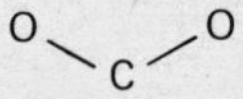

Structural formula for carbon dioxide

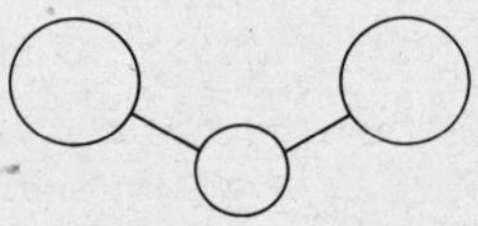

Carbon dioxide model

O—O

Oxygen molecule

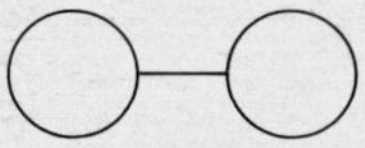

Oxygen model

(8) What energy and chemical must be added to your photosynthesis equation? ______

____

(9) What are the raw materials of photosynthesis? ______________________

____

(10) What are the products of photosynthesis? ______________________

____

(11) Why is cellular respiration called the "fire" of life? ______________________

____

____

NAME DATE

(12) In what chemical molecule is the energy of respiration stored? ______________________

____

Investigation

(A) Read the directions carefully on your Procedure page. In this activity, you will determine the rate of photosynthesis by recording the number of oxygen bubbles escaping from the cut end of *Anacharis*.

(13) Why was sodium bicarbonate added to each test tube? ______________________

____

(14) What do the bubbles tell you about the rate of photosynthesis? ______________________

____

(15) Under which condition is the bubble count the greatest? ______________________

____

(16) Under which light condition is the bubble count the least? ______________________

____

(17) What conclusions can you draw from this Investigation? ______________________

____

____

Use the Table below to record your observations.

Environmental Condition	Rate per Minute	Rate per 5 Minutes	Average Rate per Minute
Sunlight or reflector lamp			
Daylight			
Diffuse light			
Darkness			

(18) How do you account for the variation in the rate of bubbles in the four plants?

____

____

____

(B) Follow the procedure for this investigation to demonstrate one of the products of cellular respiration.

(19) Why did you leave one test tube without *Anacharis*? ______

(20) Which of the three tubes showed a change in color? ______

(21) Why did this color change happen? ______

(22) What process do you think has taken place? ______

Summary

(1) Describe the parts of the ATP molecule. ______

(2) How is the energy from ATP released? ______

(3) Write a balanced equation for cellular respiration. ______

(4) Write a balanced equation for photosynthesis. ______

(5) List the four things necessary for photosynthesis to take place. ______

(6) How can the rate at which photosynthesis takes place be determined? ______

(7) Under which light condition was the bubble count the greatest? ______

(8) What can you conclude about the rate of photosynthesis and varying intensities of light? ______

(9) What evidence did you find that respiration occurs in *Anacharis*? ______

NAME DATE

7 DNA and Its Structure

MATERIALS

INVESTIGATION▶

- construction paper
- cardboard
- tracing paper
- scissors
- glue or tape
- colored pencils
- toothpicks, flat edge
- fine marking pen

OBJECTIVES▶

- To understand the basic structure of the DNA molecule.
- To construct a model of the DNA molecule.
- To identify and label the molecular structure of DNA.

Problem

What is DNA? What is the structure of DNA?

DNA, the abbreviation for *d*eoxyribo*n*ucleic *a*cid, is sometimes called the "blueprint of life." This is because the DNA molecule is responsible for hereditary traits and protein synthesis. It also controls all cell activities in living things. This marvelous chemical is the substance of genes which are located on chromosomes in cell nuclei.

The basic unit of DNA is called a *nucleotide*, containing a deoxyribose (sugar) molecule, a phosphate, and a base. In the early 1950's, James Watson and Francis Crick proposed a model of the DNA molecule in the shape of a double helix — a twisted, spiral ladder. This tremendous discovery has helped scientists to understand how DNA controls protein synthesis, hereditary traits, and other cell activities.

Procedure

Part I. Analysis

DNA was first linked with hereditary traits during a series of experiments on two kinds of pneumococcus bacteria. The bacteria which cause pneumonia have capsules surrounding themselves; the bacteria without capsules do not cause pneumonia.

Study the diagram of the following experiment on your Data Sheet page. Biologists extracted nucleoprotein (the substance believed to control heredity at that time) from the harmful bacteria. Then they separated the nucleoprotein into DNA and protein and added each to separate cultures of harmless bacteria. The protein substance added to the harmless bacteria produced no change. However, the DNA from the harmful bacteria caused a *transformation* in the harmless bacteria culture! The harmless (without capsule) bacteria were transformed into harmful (with capsule) bacteria. In addition, they passed on this trait to their offspring. DNA was found to be the transforming factor. Report your conclusions by answering questions 1–7.

Part II. Investigation

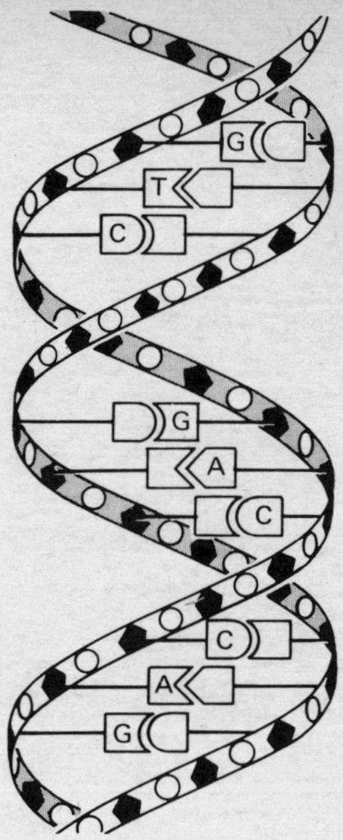

(A) On your Data Sheet, look at the structural formulas and models of deoxyribose, phosphoric acid, and the four bases — guanine, thymine, cytosine, and adenine. These are the building blocks of DNA.

The DNA molecule is in the shape of a double helix. A helix is a spiral shape. The double helix forms the sides of the molecule or "twisted ladder" as shown.

The sides are composed of alternating units of deoxyribose sugar and phosphate. The internal building blocks of the molecule, or rungs of the ladder, are composed of nitrogen-containing bases that are joined to the sides. The basic unit of DNA is called a *nucleotide*, consisting of a deoxyribose molecule, a phosphate, and a base.

There are four types of bases in DNA. Two of the bases are *purines*. The purines are *adenine* and *guanine*. The other two bases are *pyrimidines*, called *thymine* and *cytosine*. The bases are known by their code letters A, G, T, and C. These bases always bond in a certain way. Adenine bonds only to thymine and guanine bonds only to cytosine. Label the helix diagram on your Data Sheet page, filling in the base code letters and the nucleotide labels. Answer questions 8–14.

(B) Construct your own DNA model. Trace the models of deoxyribose, phosphoric acid, and the nitrogen bases pictured on your Data Sheet. Cut out the following number of each molecular model:

14 deoxyribose
12 phosphoric acid
4 of each base

Choose one of the two DNA models (straight or twisted ladder) illustrated in the Data Sheet. If you choose the "straight" DNA model, use toothpicks to attach the phosphate to the deoxyribose (toothpicks represent the carbon-oxygen bonds). If you choose the "twisted ladder" model, cut out S-shaped pieces of construction paper (wide enough to contain deoxyribose and phosphate on the ladder sides).

Then glue your model pieces on cardboard using toothpicks as carbon-nitrogen bonds between each deoxyribose and nitrogen base. Use colored pencils to distinguish your molecular pieces. Then label each model part with a marking pen.

Investigations On Your Own

(1) Make another DNA model using beads, gum drops, plastic, or jelly beans. Try to make your model three-dimensional.
(2) Study the history of the discovery of DNA. Although Watson and Crick proposed the DNA model, many other scientists were involved in this discovery. Find out the range of DNA research now being investigated.

NAME DATE

7

DNA and Its Structure

DATA SHEET

Analysis

Read your Procedure page to understand the following experiment. If you were the scientist and these were your findings (as illustrated below), what would be your conclusions about DNA?

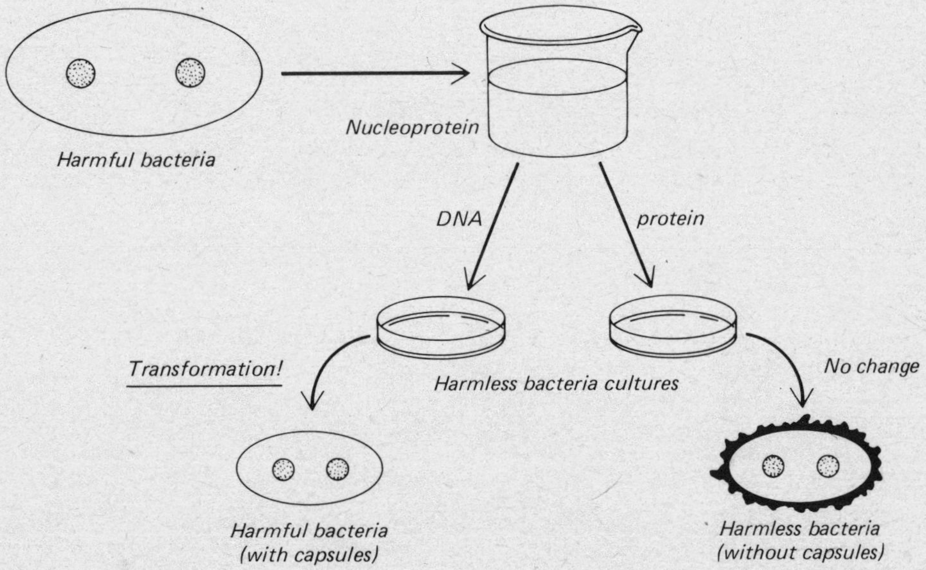

(1) What is the difference between the two kinds of pneumococcus bacteria? ______

______

______

(2) What substance did scientists extract (take from) the harmful bacteria? ______

______

(3) Why was it important to separate nucleoprotein into DNA and protein? ______

______

(4) What happened to the culture of harmless bacteria when protein was added?

______

(5) . . . when DNA was added? ______

______

(6) What kind of bacteria did the culture with DNA produce? ______

(7) What does DNA control? ______

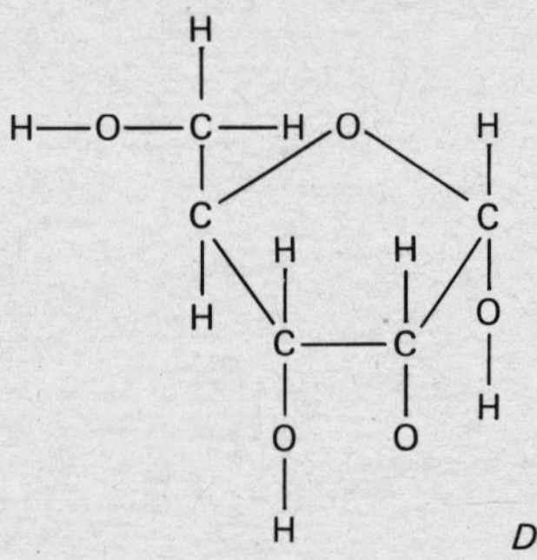

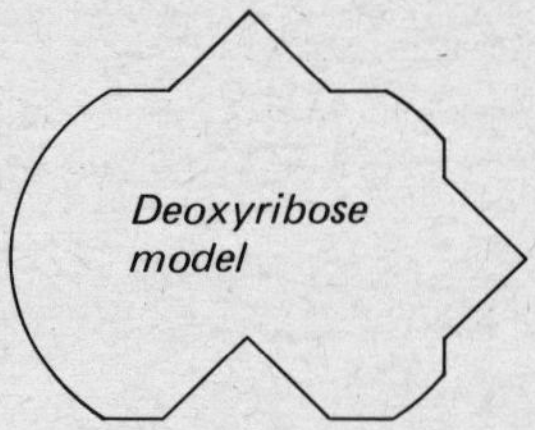

Deoxyribose

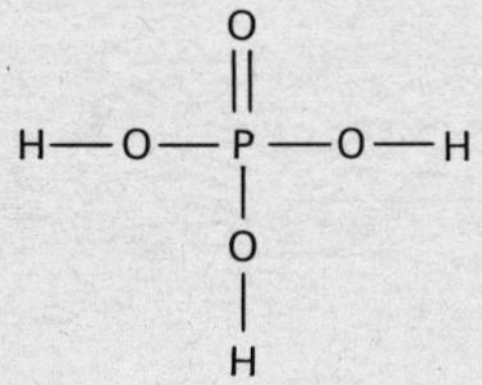

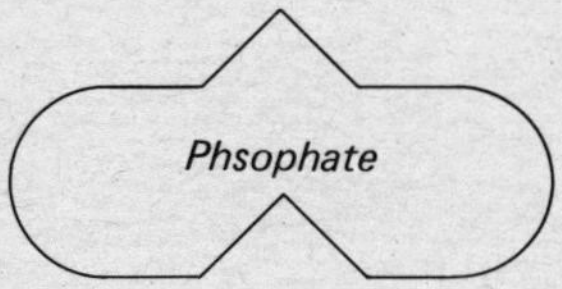

Structural formula
of phosphate

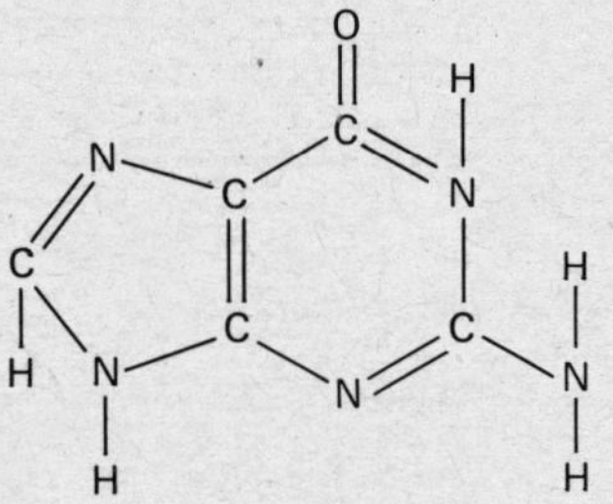

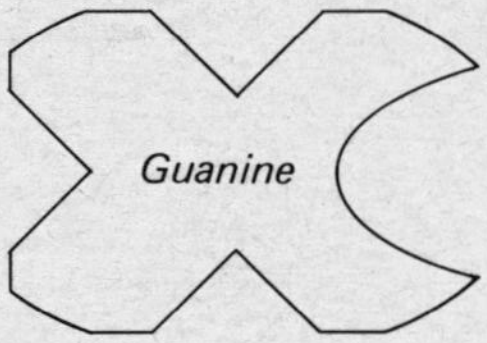

Guanine base

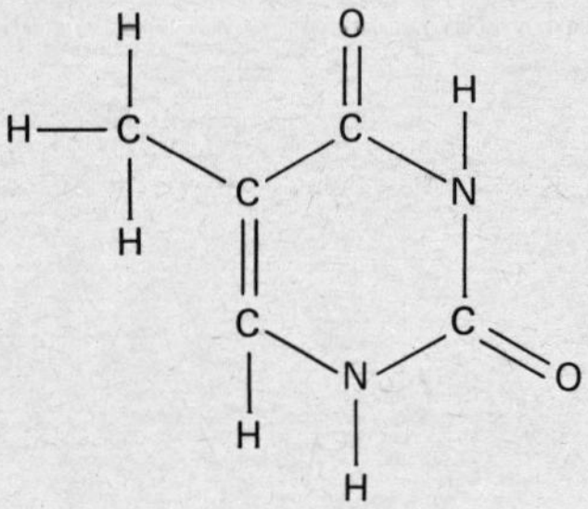

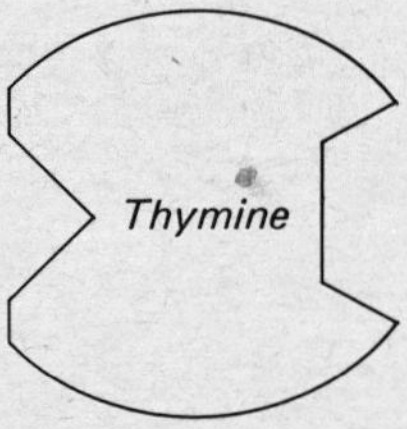

Thymine

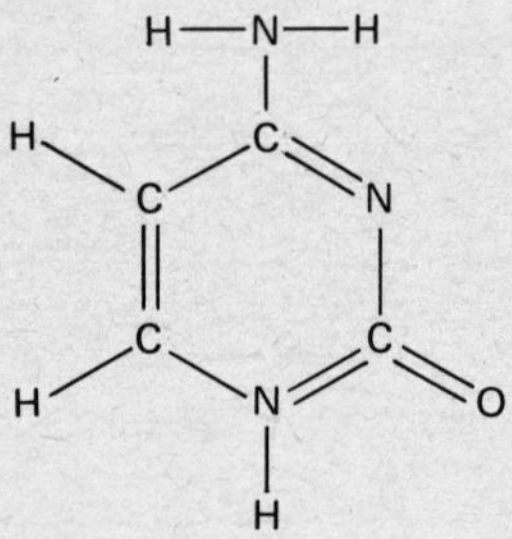

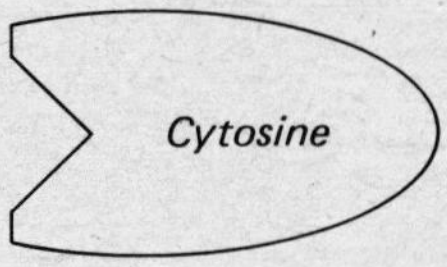

Cytosine base

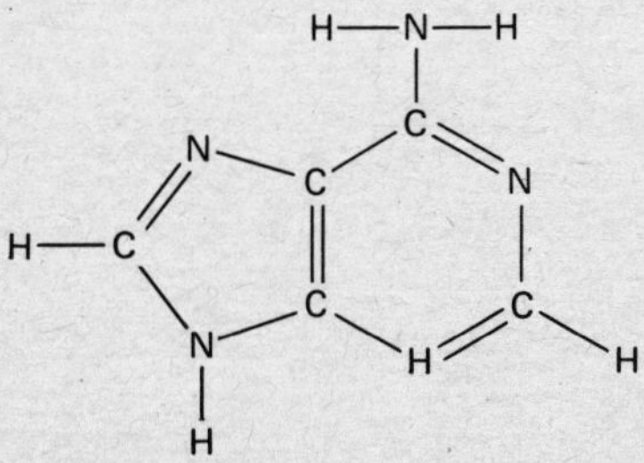

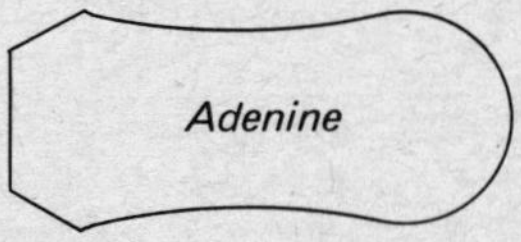

Adenine base

NAME DATE

Investigation

(A) In the diagram below, label each part of the helix and fill in the base code letters that correspond to the correct nucleotide. Be sure to label: phosphate unit, deoxyribose, carbon-nitrogen bond, carbon-oxygen bond, guanine, cytosine, thymine, and adenine.

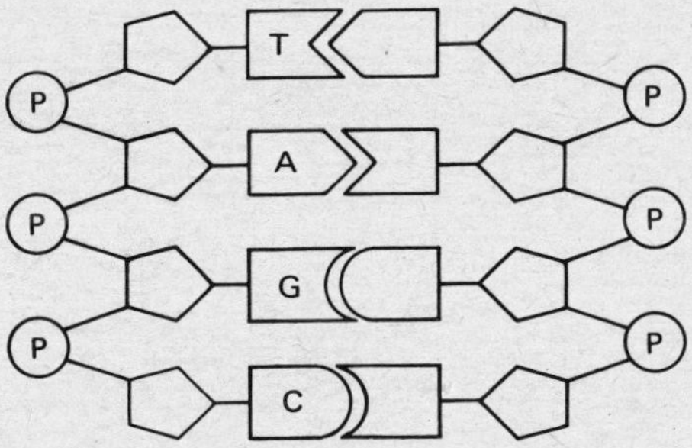

(8) Given the following code letters, predict to which base each would be bonded.

A G T C T T C G A G T A C G G

(9) What is the general structure of the DNA molecule? ______________________

____

(10) What comprises the backbone or side pieces of the DNA molecule? ______________________

____

(11) What are the three basic units or molecules of DNA? ______________________

____

(12) What is this three-part unit called? ______________________

(13) What are the purine bases of DNA? ______________________

. . . the pyrimidine bases of DNA? ______________________

(14) In what sequence do these bases bond together? ______________________

____

(B) The DNA molecule is actually in the shape of a double helix or "twisted ladder." But if a portion of the molecule is unwound, it is sometimes easier to see the parts within DNA. Choose one of the two models pictured below to make a small part of a DNA molecule. Refer to the Procedure page for specific directions.

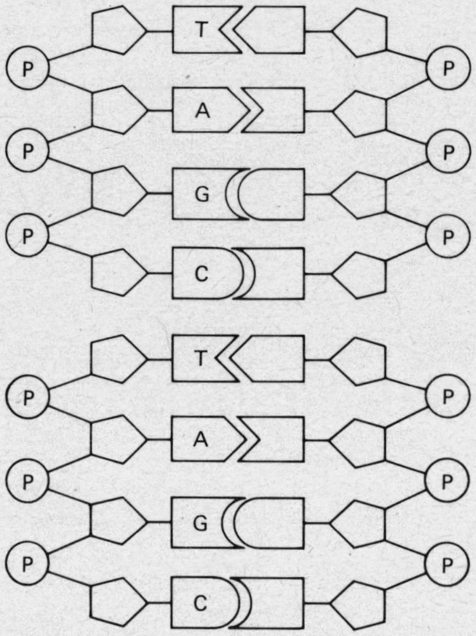

"Unwound" DNA

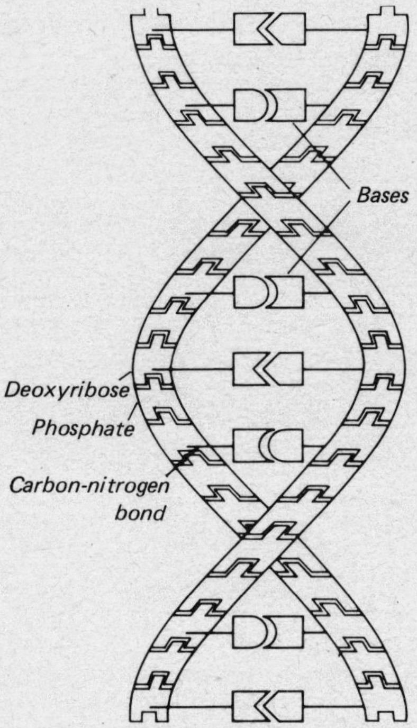

DNA "twisted ladder"

Summary

(1) Why is DNA called the "blueprint of life"? ______________________________

(2) Why is it important for scientists to know the structure of DNA? ______________________________

(3) What molecules make up the ladder sides of DNA? ______________________________

(4) What are the internal blocks or rungs of the DNA ladder? ______________________________

(5) What is the basic unit of DNA? ______________________________

(6) Name the molecules in the basic unit of DNA? ______________________________

NAME DATE

Mitosis and Meiosis 8

MATERIALS

INVESTIGATION▶
- prepared slide of onion root tip
- microscope
- prepared slide of whitefish blastula
- colored pencils

OBJECTIVES▶
- To recognize the different stages of mitosis
- To examine the mitotic phases in plant and animal cells
- To diagram chromosome changes in meiosis

Problem

What occurs during the different phases of mitosis? Why is the cell division called meiosis necessary for the production of sex cells (gametes)?

Living things must grow in order to survive. Organisms grow by the process of cell division known as mitosis. In mitosis, the nucleus of the cell undergoes a series of changes resulting in two nuclei. Then the cytoplasm divides and two new cells (identical to the parent cell) are formed. The phases of mitosis are called *prophase, metaphase, anaphase,* and *telophase*. The period between phases is called *interphase* — a "resting stage" when the cell is busy with activities other than cell division.

Meiosis is a special kind of cell division reserved for the sex cells — the female egg cell and the male sperm cell. There are two stages in meiosis with cell division taking place in each stage. The resulting gametes have half, or haploid number of, the chromosomes. When fertilization — the union of sperm and egg — takes place, the new individual has the diploid number of chromosomes.

Procedure

Part I. Analysis

Locate the phases of mitosis illustrated in the diagram on your Data Sheet page. Note that the cells are in various stages of dividing. Refer to your text to help you identify and label the various stages. Also label the nucleus, nucleolus, spindle, chromatid, and cell plate. Then answer questions 1–4 on your Data Sheet.

Part II. Investigation

(A) Obtain a prepared slide of an onion root tip. Examine your slide under the low power of your microscope.

In the onion root tip, most of the dividing cells will be found between the tip of the root and the place where the cells are beginning to elongate. Observe the cells more closely under high power. Try to locate as many phases of mitosis as you can on your slide.

Compare the mitosis in the onion root with the diagram in Analysis.

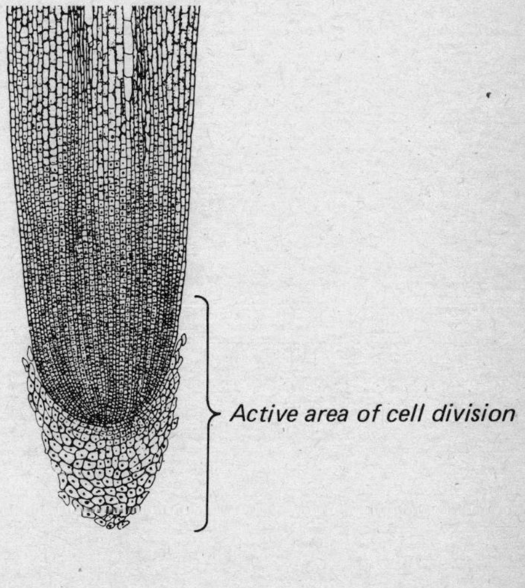

Now observe cell division in a prepared slide of a whitefish blastula. Remember to focus and center the slide first under the low-power objective. Then switch to the high-power objective to observe more details. Try to locate as many phases of mitosis as you can on your slide. Can you see differences between the plant and animal slides? Draw animal cells in each stage of mitosis in the circular spaces provided on the Data Sheet. Answer questions 5–8.

(B) All living cells divide by the process of mitosis — with the important exception of reproductive cells called gametes (or sex cells). In sexual reproduction, a male and female gamete unite during fertilization and a new individual is produced.

Look at the drawing below of human male and female chromosome pairs. Each has 23 pairs of chromosomes or 46 chromosomes — the full, or diploid, number of chromosomes. But the male sex cell (sperm) and the female sex cell (egg) have half, or the haploid number, of the chromosomes. The special kind of cell division that results in the haploid number of chromosomes is called *meiosis*. This guarantees that the new individual (formed by the union of egg and sperm) will have the full or diploid number of chromosomes.

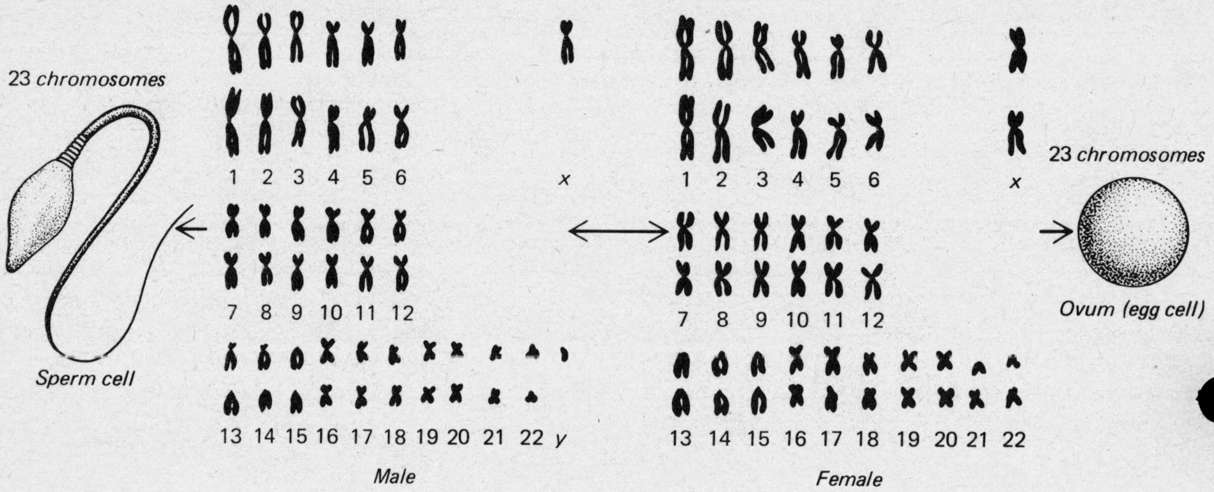

On your Data Sheet, fill in the simplified diagrams of egg and sperm formation to indicate changes that occur during meiosis. Refer to your text to help you identify the stages. Draw two pairs of chromosomes (use a different color for each) in the first stage of egg and sperm formation. Follow the steps in the diagram by filling in the chromosome pair(s). Then answer questions 9–15.

Investigations On Your Own

The union of the sperm and egg cells is called fertilization. Illustrate the chromosome makeup of these cells by using colored pencils to fill in the diagram provided. When the zygote (fertilized egg) divides, body cells are formed. Fill in the chromosomes of the body cells. *The normal, diploid number of chromosomes for this individual is 2 pairs or 4 chromosomes.*

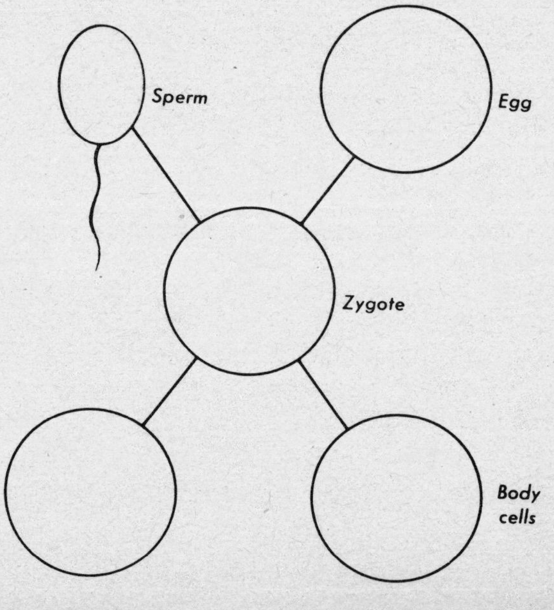

NAME DATE

(B) Fill in the chromosome pairs to show the changes during meiosis.

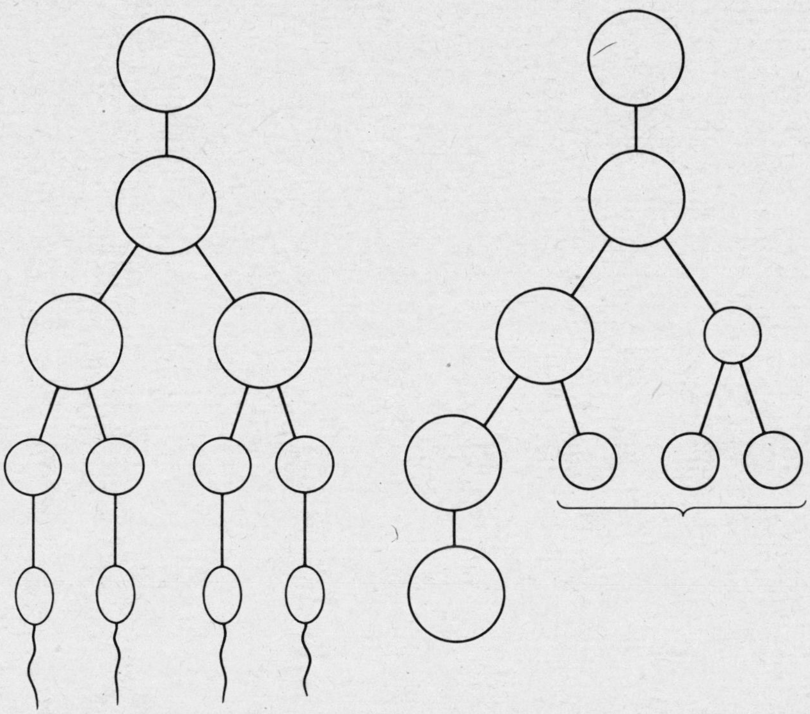

(9) What happens in the primary oocyte and the primary spermatocyte? ______

______

(10) What are the chromosomes called following this duplication? ______

______

(11) What happens between the primary and secondary stages? ______

______

(12) What happens when the chromatids separate during the division of the secondary oocyte and spermatocyte? ______

______

(13) What is the chromosome number of the ootid and the spermatid? ______

______

(14) What happens to the ootid? ______

to the spermatid? ______

(15) What is the function of the polar bodies in reproduction? ______

______

Summary

(1) List the stages of mitosis.

(2) What is the significance of the mitotic process?

(3) How does mitosis in plant cells differ from mitosis in animal cells?

(4) What structures are present in both plant and animal cells?

(5) In which cells does meiosis occur?

(6) How would you define meiosis?

(7) What is the significance of meiosis?

(8) During which stages of the reproductive process does meiosis occur?

(9) What would occur if there were no reduction of chromosome number?

NAME DATE

Principles of Heredity 9

MATERIALS

ANALYSIS▶
- colored pencils

INVESTIGATION▶
- 28 ice cream sticks
- adhesive tape
- scissors
- marking pens, fine point
- construction paper

OBJECTIVES▶
- To diagram the inheritance of traits in monohybrid and dihybrid crosses by completing Punnett squares
- To explain incomplete dominance
- To understand the role of chance in the inheritance of traits
- To construct a gene-chromosome model for the garden pea plant

Problem

How are traits inherited from one generation to the next? What is the role of chance in the inheritance of traits?

Inherited traits are determined by chemical factors called genes that are passed from parents to offspring. Genes are located on chromosomes within cell nuclei. Usually genes occur in pairs so that an individual has two genes for each trait. If the genes for a particular trait are alike, the individual is homozygous for that trait; if the genes are different, the individual is heterozygous or hybrid. Each parent sex cell contributes half the genetic material to the new offspring. Since each parent actually has one of two genes to offer, the law of chance plays a part in the study of genetics.

Procedure

Part I. Analysis

(A) One of the traits studied by Mendel was flower position on the garden pea plant. *Axial flowers* grow along the stem. *Terminal flowers* develop only at the tip of the stem. Mendel found that axial (*A*) flower position is dominant to terminal (*a*) flower position. Diagram the inheritance of flower position when the female plant passes on the recessive gene *a* for terminal flowers, and the male passes on the dominant gene *A* for axial flowers. Fill in the Punnett square to find out the offspring in the first generation — F_1. Then diagram the results of a cross between two F_1 parents. Answer questions 1–9 on the Data Sheet.

(B) In horses, black coat *B* and trotting gait *T* are dominant genes while the corresponding genes (or alleles) are the recessive genes — white coat *b* and pacing gait *t*. When two homozygous horses of these opposite traits are crossed, the first generation is hybrid for both traits. See the diagram below.

The letters BBTT, bbtt, and BbTt, represent the genotype or gene makeup. The picture shows the phenotype or physical expression of the trait. Note that if the parents are homozygous for two traits, they can only contribute one gene for each trait. Answer questions 10–12 on the Data Sheet.

Black Trotter BBTT × White Pacer bbtt
↓
Black Trotter BbTt

Complete the Punnett square on your Data Sheet in the space provided to find out the probable offspring of two F_1 parents. Then answer questions 13 and 14 on your Data Sheet. Remember that a Punnett square is helpful when you predict gene combinations and estimate the probable percentages of expected results.

Part II. Investigation

Make a gene-chromosome model for a garden pea plant. Cut out a 30-cm circle from construction paper to represent the cell nucleus in a pea plant. Attach a piece of adhesive tape to each of the 28 ice cream sticks. Use a fine pen to label the taped sticks. Label 14 sticks with the dominant traits and 14 sticks with the recessive traits. There should be two sticks for each dominant trait and two sticks for each recessive trait. Do not attach the "chromosome" ice cream sticks to the "nucleus." This will enable you to interchange genetic models.

DOMINANT TRAITS	*RECESSIVE TRAITS*
tall stem	short stem
axial flowers	terminal flowers
round seed	wrinkled seed
yellow seed	green seed
colored seed coat	white seed coat
inflated pod	constricted pod
green pod	yellow pod

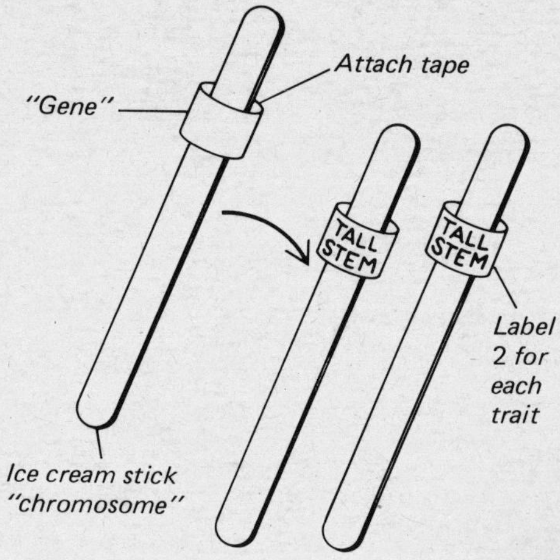

The pea plant nucleus contains 7 pairs of chromosomes or 14 chromosomes. Your label on the chromosome ice cream stick represents the dominant or recessive gene. Each pea plant trait studied by Mendel is located on different chromosome pairs. Using your Model, answer questions 15–17 on the Data Sheet.

Randomly arrange the 7 dominant pairs of traits in the "nucleus." Be sure to place each chromosome pair for a particular trait next to one another. Answer questions 18 and 19 on the Data Sheet. Then, in the space provided, draw your gene-chromosome model, and draw the physical appearance of such a pea plant.

Replace one chromosome per pair with the chromosome containing the recessive trait or gene. Answer questions 20 and 21 on the Data Sheet. Then change your gene-chromosome model by replacing all dominant chromosomes with recessive ones. Answer questions 22 and 23.

Investigations On Your Own

Using your imagination, create other possible combinations of genes with your chromosome sticks. Remember to keep chromosome pairs for the same trait together.

NAME DATE

Principles of Heredity 9

DATA SHEET

Analysis

(A) Fill in the Punnett square to find the first generation from a cross of a male axial (AA) plant and a female terminal (aa) plant. Then complete the Punnett diagram of an F_1 cross to determine the second generation.

P ♂ axial (AA) x ♀ terminal (aa)

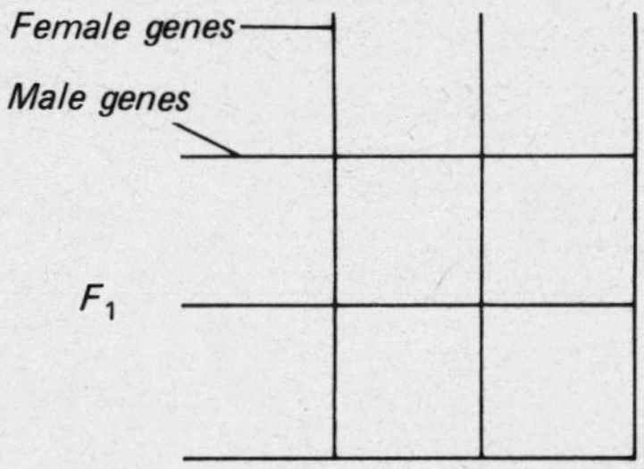

F_1 Axial (Aa) x axial (Aa)

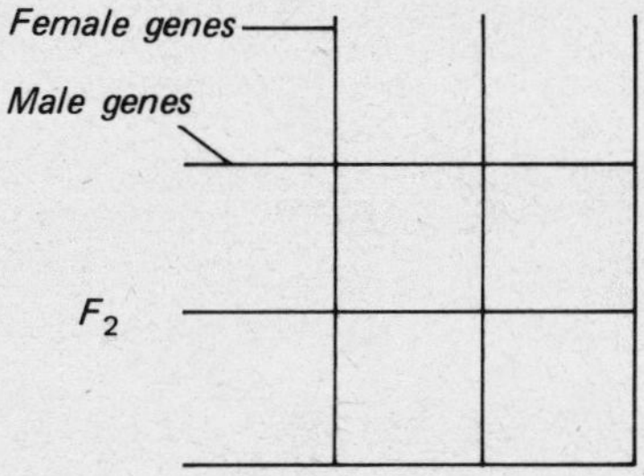

(1) Which parent plant is homozygous for the dominant trait? ________________

for the recessive trait? ________________

(2) Name the gene contribution of the male parent. ________________

of the female parent. ________________

(3) Which genes are inherited by the F_1 offspring? ________________

(4) What will be the position of the flowers produced by the F_1 plants? ________________

____

(5) How does this cross demonstrate the Principle of Dominance? ________________

____

____

(6) Explain the Principle of Unit Characters. ________________________________

____

(7) Why is the cross of the first generation offspring called a monohybrid cross?

____

____

(8) What ratio do you predict for phenotype among the F_2 offspring? ____________

____

(9) What is the predicted ratio of genotype among the F_2 offspring? ____________

____

(B)

(10) What is the genotype of the male? ____________ the female? ____________

(11) What genes are present in the sperms? ____________ the eggs? ____________

(12) What is the genotype of the F_1 generation? ____________________

the phenotype of the F_1 generation? ____________________

Diagram the possible offspring from a cross of two F_1 parents in the Punnett square below. Then list the ratio of phenotypes.

____________ black-trotters

____________ black-pacers

____________ white-trotters

____________ white-pacers

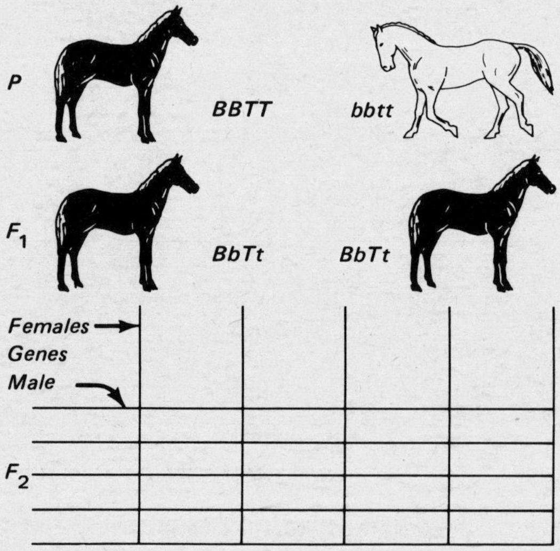

(13) In the spaces below, summarize the genotypes of the F_2 offspring.

____________	BBTT	____________	BbTT	____________	bbTT
____________	BBTt	____________	BbTt	____________	bbTt
____________	BBtt	____________	Bbtt	____________	bbtt

NAME DATE

(14) Are the ratios obtained from these offspring predictable? Explain. ____________________

____

____

Investigation

(15) What do the paired ice cream sticks represent? ____________________

____

(16) What does each labeled tape represent? ____________________

(17) How could you represent other traits (genes) in your model? ____________________

____

(18) What is the genotype of this model? ____________________

(19) Describe the phenotype (physical characteristics) that would result from your gene-chromosome model. ____________________

____

____

In the circular space, draw your gene-chromosome model. Then draw or describe the physical appearance of such a pea plant in the space provided.

(20) How does this model differ from the first model? ____________________

____

____

(21) In what way is this model similar to the first model? ____________________

____

(22) Describe the genotype. ____________________

(23) Describe the phenotype based on this particular model ______________________

____

____

____

Summary

(1) Briefly define Mendel's Law and the Principles that have been demonstrated in the crosses you have completed.

Principle of Dominance ______________________________________

____

Principle of Unit Characters ______________________________________

____

Law of Independent Assortment ______________________________________

____

____

(2) In a cross of two individuals heterozygous for a single trait, what genotype and phenotype ratios would be obtained? ______________________________

____

(3) What phenotype ratio is obtained when individuals heterozygous for two traits are crossed? ______________________________________

(4) Explain if the Law of Independent Assortment would apply if the genes for coat color and gait in horses were carried on the same chromosomes. ______________

____

____

(5) On the basis of your results, explain how genetic principles yield predictable results. ______________________________________

____

NAME DATE

Sex-Linked Traits 10

MATERIALS

ANALYSIS▶
- 2 pennies, adhesive tape

INVESTIGATION▶
- cotton plugs for jars
- etherizing jar
- cotton wad
- adhesive tape
- ethyl-ether
- caps for jars
- grease marking pencil
- 2 fresh culture jars (large baby food jars)
- culture of wild-type fruit flies
- culture of fruit flies (red-eyed males and females)
- culture of fruit flies (white-eyed males and females)
- hand lens
- camel hair brush or toothpicks

OBJECTIVES▶
- To study how sex is determined
- To diagram the inheritance of sex-linked traits
- To become familiar with techniques used for culturing and examining fruit flies
- To study sex linkage as observed in fruit flies

Problem

What are sex-linked traits? How are fruit flies used to study inheritance?

The sex of an organism is determined by a pair of chromosomes called the sex chromosomes. The female chromosome is the *X* chromosome, the male chromosome is the *Y* chromosome. Females have two *X* chromosomes, males have one *X* and one *Y* chromosome. When the gene for a particular trait is located on the *X* chromosome, the trait is sex-linked.

The common fruit fly, *Drosophila melanogaster,* is helpful in studying inheritance. Drosophila is small and reproduces rapidly. The sex of each fruit fly, and other characteristics, can be easily distinguished.

Procedure

Part I. Analysis

(A) Each human body cell contains 23 pairs of chromosomes. One of these pairs (the sex chromosomes) differs in the male and the female. Examine the genotypes in the Punnett square diagram on your Data Sheet. Fill in the diagram of a cross between a male and a female. Answer questions 1–5.

Demonstrate the probability involved in sex determination. Tape an "X" on one side of a penny and, on the other side, tape a "Y." On another penny, tape an "X" on both sides. Now flip both coins. Note what "genes" appear. Continue the flips at least 30 times. Record the chromosome combination that appears each time on the Data Sheet Table. Complete questions 6–9.

(B) When a gene is carried on a sex chromosome, it is said to be sex-linked. In humans, the genes controlling color vision are located on the *X* chromosome. The *X* chromosome carrying a gene for normal vision is represented by X^C while X^c represents a gene for color blindness. Study the Genotypes Chart, on your Data Sheet, which shows the female and male genotypes of varying vision abilities. Use the Punnett square to determine the color vision ability of the offspring of a female (X^CX^c) and a male (X^CY). Using the results of your diagram and the information on the genotype chart, answer questions 9–13.

Part II. Investigation

(A) Obtain a jar in which wild-type fruit flies have been cultured. To examine these flies, it is necessary to immobilize them through the process of etherization. Prepare an etherization jar and a re-etherization device as shown in the figure below.

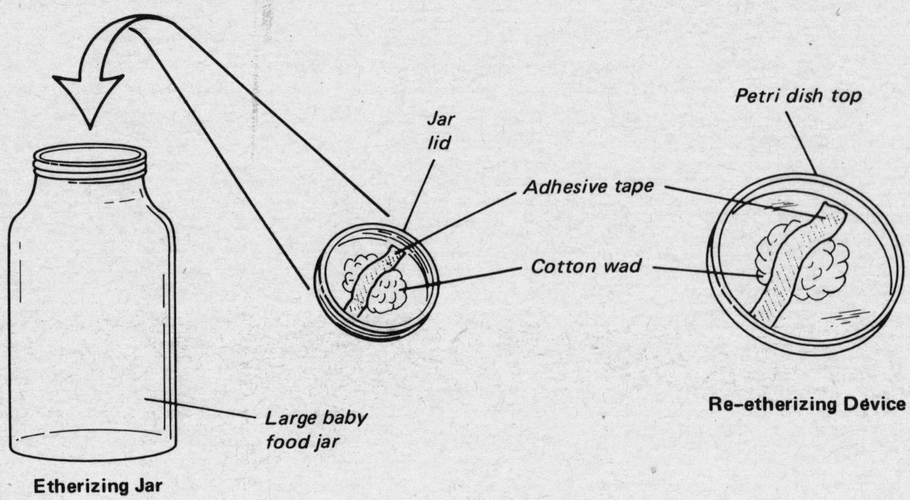

Carefully read each of the following steps before beginning the procedure.

Procedure for Etherization and Examination of Fruit Flies

1. Place several drops of ethyl-ether on the cotton wad of the etherizing jar and temporarily replace the lid. (**Warning:** Ether is a highly volatile and toxic substance. USE WITH CAUTION.)

2. Tap the culture jar gently to dislodge the flies from near the mouth of the culture jar.

3. Remove the cotton plug from the culture jar and the lid from the etherizer. Quickly bring the opened mouths of the jars together.

4. Invert the jars so that the culture jar is on top. Hold the jars firmly together.

5. Tap the culture jar to cause the flies to fall into the etherizer. Recap the culture jar.

6. After about 10 seconds, pour the etherized flies out on a sheet of white paper. Examine their body characteristics.

7. Recovery of flies takes about 5 minutes. You must examine the flies quickly. Under-etherized flies will recover in about one minute. To re-etherize them, place a few drops of ethyl-ether on the small cotton wad attached to the lid of a Petri dish. Place this re-etherizer over the awakening flies.

Use the instructional drawing on the next page as a guide. Answer questions 14–17.

NAME DATE

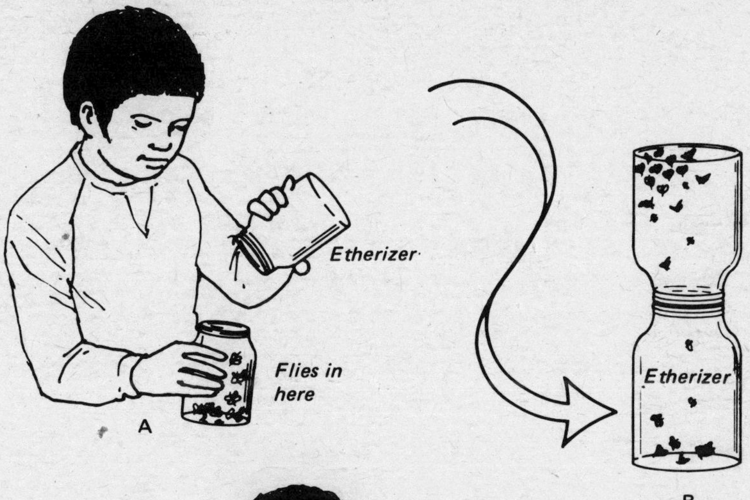

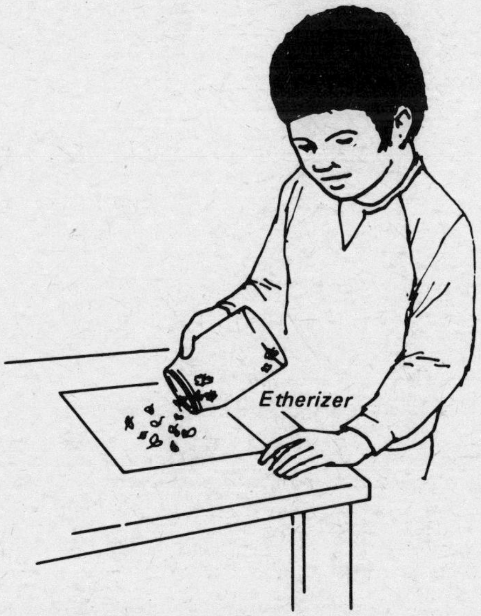

(B) Examine traits shown in the drawings below and the chart on your Data Sheet. Try to become familiar with some of the major characteristics that may be observed in fruit flies. As you examine the flies with a hand lens, use a camel hair brush or toothpick to separate and classify the flies according to the characteristics observed. As your examination proceeds, you should be able to determine that some have a broader abdomen than others. These are the females. The smaller males have a black-tipped, blunt, posterior end. Record your observations on the Data Sheet Table.

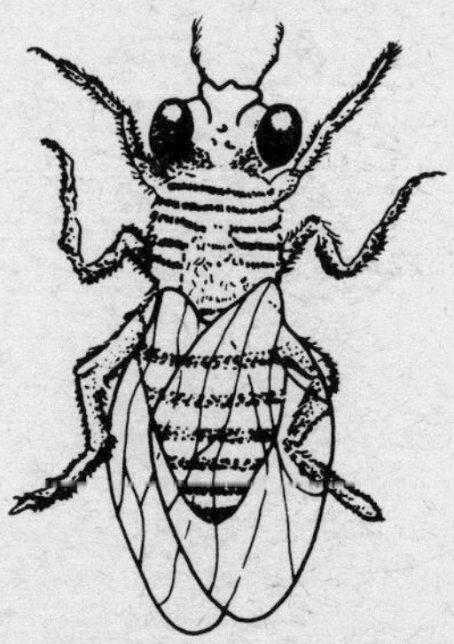

Adult female ♀

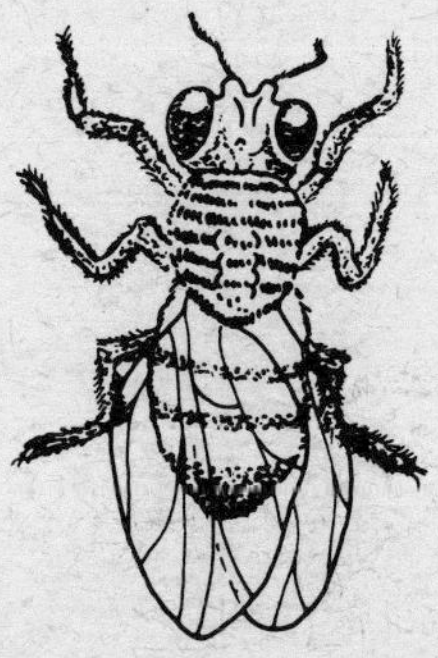

Adult male ♂

(C) Experimental mating should be made with females having unfertilized eggs. Virgin females may be obtained by removing all adults from the cultures of the red-eyed and white-eyed flies. Examine these cultures 10–12 hours later to see if new adults have emerged. These flies should not have mated and may be used in the matings to be performed. Etherize the new adults and select from the culture 2 or more red-eyed males, and 2 or more red-eyed virgin females. Do the same with the white-eyed culture. In a new culture jar, place 2 or more white-eyed males and 2 or more red-eyed virgin females.

In another culture jar, place 2 red-eyed males and 2 white-eyed virgin females. Identify each culture bottle with the parent cross and date. Set them aside in a location at room temperature and away from bright light. After a week, examine the cultures. Answer questions 18–21.

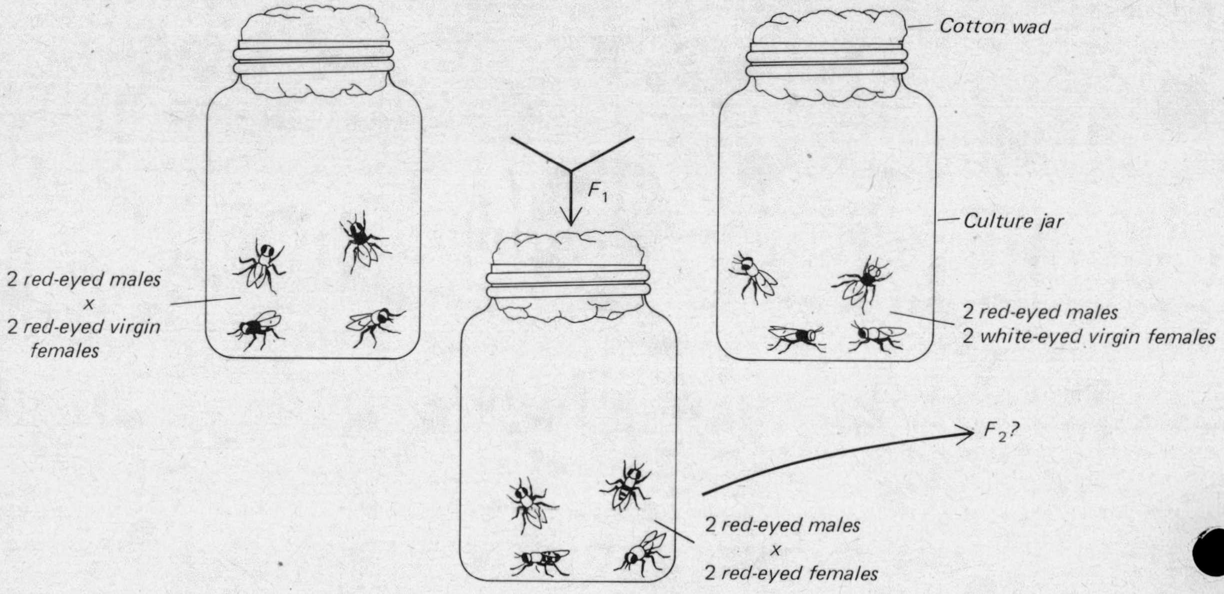

Remove and discard the parent flies. Examine the cultures daily for the emergence of the F_1 generation flies. As emergence progresses, etherize and determine the sex and eye color for each of the flies. Continue etherization and examination of the F_1 generation until at least 100 flies have been classified. *Do not return the flies to the original culture.* Record your findings of the F_1 generation in the Data Sheet Table. Study the F_1 generation for 7 days.

From the F_1 generation offspring, select 2 or more red-eyed males and 2 or more red-eyed females and place them in fresh culture jars. After 7–8 days, examine the cultures. Remove the parent flies. Etherize as before and identify the eye color and sex of the F_2 offspring as they emerge. Record your data in the F_2 Table.

Investigations On Your Own

Hemophilia is a sex-linked trait of humans inherited in the same manner as color blindness. Because it often results in death, it is said to be lethal. Prepare a report on this sex-linked disease and other sex-linked traits not studied in this Investigation. Include in your report an analysis of where it occurs in populations and the consequences of it being inherited.

NAME DATE

Sex-Linked Traits 10

DATA SHEET

Analysis

(A) Fill in the Punnett square diagram below and then answer the questions that follow.

♀XY X ♂XX

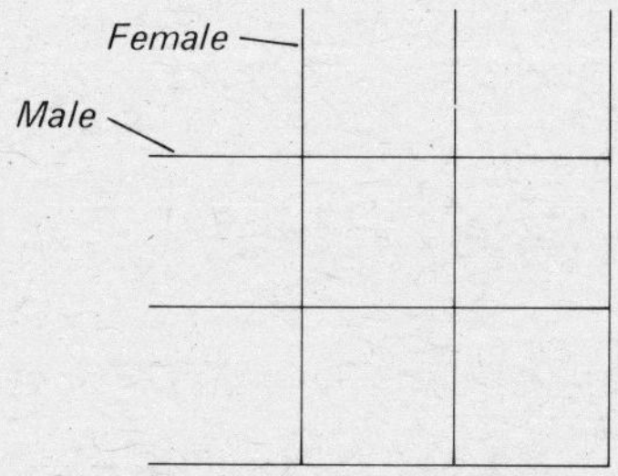

(1) What is the 23rd pair of chromosomes called? ____________________

(2) What is the genotype of the female ♀ ? ____________________

(3) Following meiosis, how many different sex chromosomes will be in the eggs?

____

(4) What difference may be seen in the genotype of the male ♂ ? ____________________

(5) What is the possibility of chromosomes in sperms following meiosis? ____________________

____

Record the chromosome combinations from flipping pennies as described on the Procedure page.

Chromosome combinations from flipping pennies

X and X
X and Y

(6) What "chromosomes" appear? ____________________

____

(7) What is the approximate ratio of genotypes obtained from flipping the pennies?

____

(8) From the Punnett square and the "penny" activity, what is the probability of male offspring? ____________________ of female offspring? ____________________

(9) Which gamete actually determines the sex of the offspring? ____________________

(B) Use the genotype chart below to help you diagram inheritance of color blindness and complete the questions.

GENOTYPES		
Female	**Vision Ability**	**Male**
X^CX^C	normal	X^CY
X^CX^c	carrier	- - - -
X^cX^c	color blind	X^cY

$X^CY \times X^CX^C$

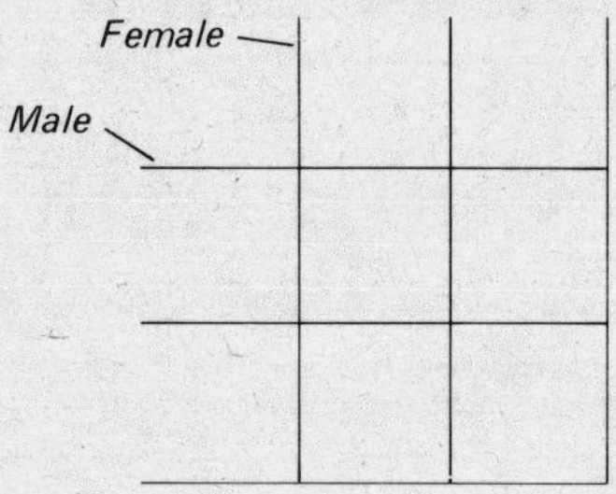

(10) What percent of the females are color blind? ____________________

(11) What is their genotype? ____________________

(12) What % of the males are color blind? __________

(13) How is it possible that some male offspring have normal vision when the male parent is color blind?

____

____

____

Investigation

(A) Follow the procedure for etherizing and examining fruit flies.

(14) Explain two reasons why it is important to bring the opened mouths of the culture jar and the etherizer together quickly. ____________________

____

____

(15) How can you detect the flies that have been over-etherized? ____________________

____

(16) How long does it take before the etherized flies recover? ____________________

____

(17) If etherized flies recover too quickly for your examination, what could you do?

____

____

____

NAME DATE

(B)

Characteristics	Dominant	Recessive
Size and Shape of Wings	long	vestigial, curved, miniature, bent
Color and Shading of Body	gray	black, yellow
Eye Color	red	brown (sepia), black, white

Study the dominant and recessive characteristics of fruit flies in the chart above. Then record the number of fruit flies you observed with those traits in the Table below.

Date	Sex		Body Color			Wings					Eye Color			
	M	F	Gray	Black	Yellow	Normal	Vestigial	Curved	Miniature	Bent	Red	White	Black	Brown

(C) Perform the experimental procedure for studying sex-linkage in fruit flies.

(18) Why must virgin females be used? ____________________

____

(19) What is the possible genotype of the red-eyed male? ____________________

of the red-eyed female? ____________________

(20) What is the genotype of the white-eyed male parent? ____________________

of the red-eyed virgin female? ____________________

(21) Give at least three reasons why fruit flies are so excellent for genetic studies.

____

____

____

Record your findings of the F_1 generation in the Table below.

	F₁ GENERATION OFFSPRING				
		Female		**Male**	
	Date	**Red-Eyed**	**White-Eyed**	**Red-Eyed**	**White-Eyed**
1.					
2.					
3.					
4.					
5.					
6.					
7.					

Use the following Table to record your findings for the F_2 generation.

F₂ GENERATION OFFSPRING				
	Female		**Male**	
Date	**Red-Eyed**	**White-Eyed**	**Red-Eyed**	**White-Eyed**
1.				
2.				
3.				
4.				
5.				
6.				
7.				

Summary

(1) On the basis of your data, discuss the body color, wings, and eye color that are dominant in the wild-type fly (as indicated in the frequency of appearance in random matings). ______________________________

____

(2) In the Investigation **(C)**, what was the purpose of two different sets of parents in obtaining the F_1 generation? ______________________________

____

NAME DATE

Human Genetic Traits 11

MATERIALS

INVESTIGATION▶ • PTC (phenylthiocarbamide papers)

OBJECTIVES▶
- To study the inheritance of some human traits
- To determine the frequency of selected traits in a given population
- To see the extent of genetic variability

Problem

What are some human genetic traits?

Human traits are difficult to study because of the relatively long human life span and the limited number of human offspring. In addition, the number of chromosome pairs (23) increases the possible genetic combinations.

However, it is possible to take a sample of a human population to find the frequency of a trait and the possible ways a given trait is inherited.

Procedure

Part I. Analysis

Geneticists probably know more about blood types than any other inherited human trait. The four basic blood types are determined by the presence or absence of the A and B agglutinogens in the red corpuscles. Consider blood types as being determined by a single pair of genes. Thus, type-A blood may be homozygous I^AI^A or heterozygous I^Ai. Type-B blood may be homozygous I^BI^B or heterozygous I^Bi. Type-O blood must be homozygous ii, and type-AB is a combination of A and B antigens, with the genotype I^AI^B, illustrating codominance.

Fill in the Punnett square diagram on your Data Sheet of a cross between a parent heterozygous for type-A blood with a parent having type-AB blood. Based on the information above, answer questions 1–5.

Part II. Investigation

(A) Read the short description of each of the following traits. Determine which trait you have (your phenotype). Record your findings in the Data Sheet Table. In recording your genotype, use only a single symbol to indicate dominance since you do not know whether you are homozygous or heterozygous for the trait.

(1) *Attached Ear Lobe:* In most people, the ear lobes hang free. But when a person is homozygous for a certain recessive gene (e), the ear lobes are attached directly to the side of the head.

(2) *Widow's Peak:* In some people, the hair line drops downward and forms a distinct point in the center of the forehead. This is known as a widow's peak. It results from the action of a certain dominant gene (W).

(3) *Tongue Rolling:* A dominant gene (R) gives some people the ability to roll the tongue into a "U" shape when the tongue is extended from the mouth. Nonrollers (r) can do no more than produce a slight downward curve of the tongue.

(4) *Bent Little Finger:* A dominant gene (B) causes the last joint of the little finger to bend inward toward the fourth finger. Lay both hands flat on the table, relax the muscles, and note whether you have a bent or a straight little finger.

(5) *PTC Tasting:* Place a piece of PTC paper on your tongue or, if you detect no taste, chew the paper. If you still detect no obvious taste, you are a nontaster and are homozygous for a recessive gene (t). The tasting of this chemical results from the presence of a dominant gene (T).

Your teacher will compile the data from the entire class or classes. Record these figures on your Data Sheet Chart.

(B) On your Data Sheet, the figure of the genetic circles permits you to visualize how much human genetic traits vary. In the center of the circle, enter the number of individuals in your class. Work from the middle to the outside. Using the data from the Table in **A**, enter the number of individuals with free ear lobes (E) in the proper space in the next circle. Enter the number of students with attached ear lobes in the proper space. Continue outward, dividing the group by each trait and indicating the number in each group. If you wish, you may extend the section by adding another trait. Answer questions 6–10.

Investigations On Your Own

Prepare a pedigree of your family's ability to roll their tongues or a pedigree of one of the other characteristics studied in this Investigation. Construct the pedigree representing two generations beginning with your parents and their offspring (which would include you). A square is used to represent a male, and a circle, a female. Shading of either square or circle indicates the possession of the recessive trait. Each generation is indicated by a Roman numeral. The squares and circles are joined by lines indicating relationships. Pedigrees may be expanded to include as many generations and relationships as are desired and available for testing. Review the pedigree shown (in outline form) to understand how relationships may be indicated.

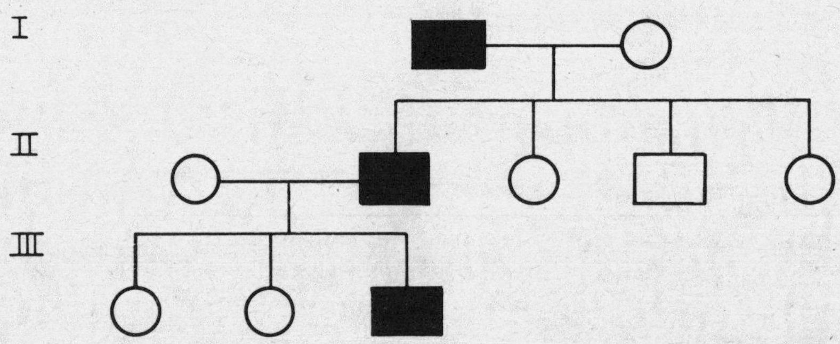

NAME ______________________ DATE ______________

11 Human Genetic Traits

DATA SHEET

Analysis

Read the information concerning blood types on your Procedure page. Then fill in the Punnett square and complete the questions that follow.

(1) What is the genotype of the parent heterozygous for type-A blood?

____

(2) What is the genotype of the AB parent?

____

(3) What genes for blood type may be passed on from the type-A parent to the offspring?

____

from the type-AB parent? ______________

(4) What gene combinations might be inherited by the offspring? ______________

____

(5) What blood types might the offspring possess? ______________

Investigation

(A) Fill in the genotype chart as described on your Procedure page.

Trait	Your Genotype	Dominant	Recessive	Ratio	Frequency (%) of Your Trait
Ear Lobes					
Widow's Peak					
Tongue Rolling					
Bent Little Finger					
PTC Tasting					

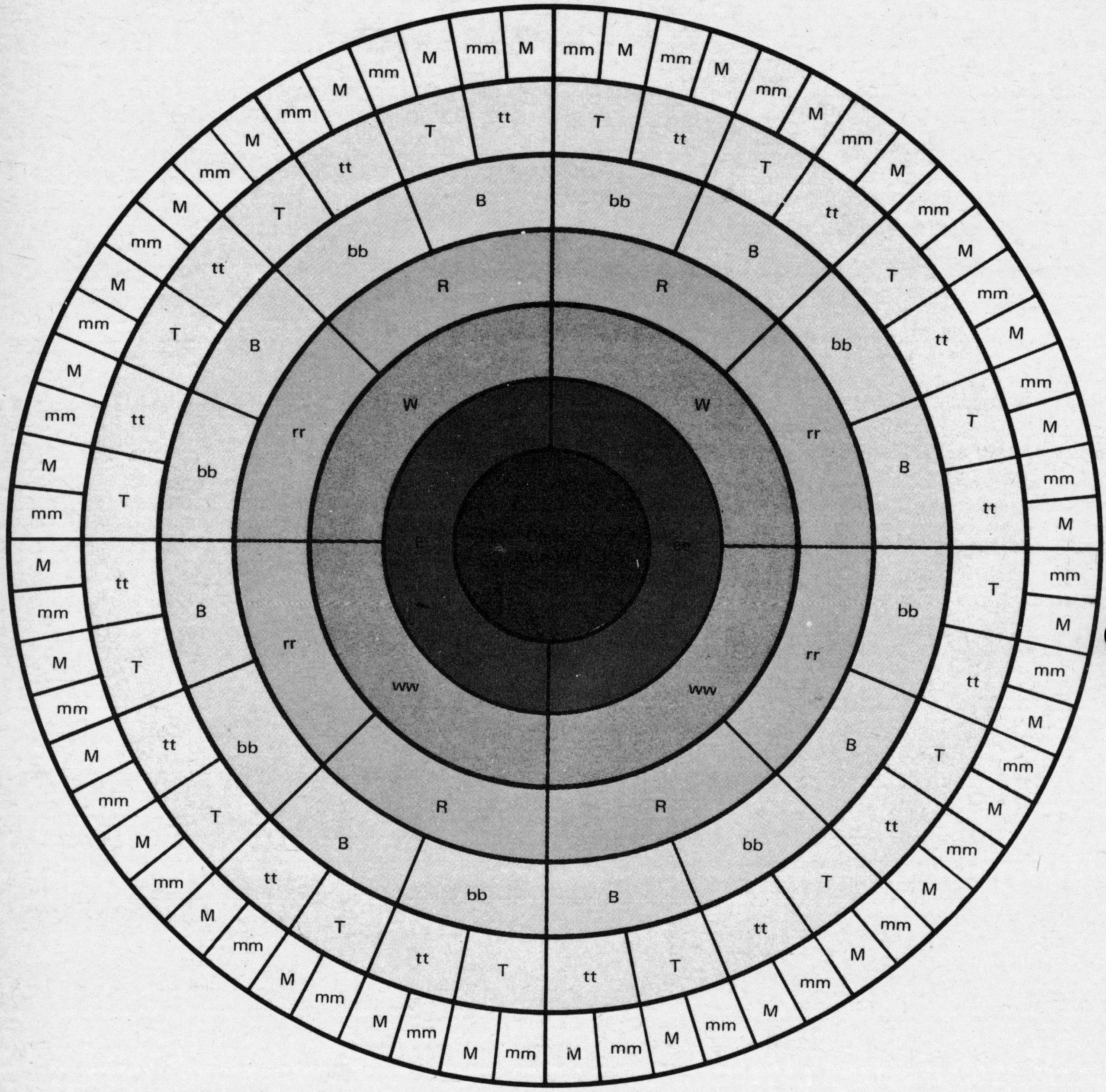
E
ee
W
W
ww
ww
R
R
rr
rr
R
R
rr
rr

NAME DATE

(B) Fill in the genetic circle chart on the opposite page.

(6) How would it help to extend the outermost section by adding another trait?

(7) If you had to keep extending the traits, how many more could you add?

(8) Why are there always two choices as you move to each step?

(9) If two people ended up on the same outer space, would they look alike?

(10) If you had 100 circles, would these two people look alike?

Summary

(1) To what extent do the ratios obtained compare to Mendelian ratios previously studied?

Explain any variation.

(2) Could two parents who are tasters have a child who is a nontaster? Explain.

(3) What situation could result in less variation in a human population than what you have observed in **(B)**?

(4) How do the ratios obtained for the characteristics observed compare to the national average? The national average is 65% for PTC tasting and tongue rolling and 35% for nontasters and nonrollers.

(5) How could you account for a variation from the national averages? ____________

____

____

____

Use the information from the Analysis section on your Procedure page to complete the following Table.

Blood type of parents	All possible genotypes of parents	All possible genotypes of children	All possible blood types of children	Blood types not possible for children
A and O				
B and O				
A and B				
AB and A				
AB and B				
AB and O				
O and O				

NAME DATE

Cloning 12

MATERIALS

INVESTIGATION▶

- Boston fern plant
- sterile distilled water
- 15% chlorine bleach solution
- liquid detergent
- 2 large containers
- forceps
- scalpel
- alcohol
- Bunsen burner
- strainer
- 150 tubes/tray Murashige Minimal Organics Medium (Nutri-Tube)
- 50 Murashige Fern Multiplication Medium (Nutri-Tube)
- fluorescent lights
- vermiculite
- plastic wrap
- planting containers
- Petri dishes

OBJECTIVES▶

- To understand the general procedure of cloning or asexual reproduction in animals and plants
- To observe the stages in plant (Boston fern) cloning
- To relate the significance of cloning to medical research on cancer, immunology, aging, and superior plant crops

Problem

What is cloning? What are the benefits of cloning plants and animals?

A clone is an exact genetic replica of a life form. The word "clone" means a twig or slip that results from asexual reproduction. Most plants and animals reproduce sexually, with hereditary material contributed by two parents. Sexual reproduction results in great diversity while cloning results in new individuals exactly like the donor plant or animal.

Cloning is used to grow commercially uniform and desirable varieties of apples, potatoes, grapes, bananas, etc., as well as numerous flowering plants (roses, orchids, carnations, etc.). Cloning of animals such as toads, frogs, and salamanders is not done for the purpose of producing uniform animals. Animal cloning is particularly helpful to medical research on cancer and aging. It is also being used to understand how cells perform different tasks and how cells build up immunities to various diseases.

Procedure

Part I. Analysis

Frogs were the first multicellular animals to be reproduced by cloning. They are ideal for cloning because of the tremendous number of eggs (about 2,000) produced per frog. In addition, fertilization and development of the embryo takes place outside the frog's body. Pond water conditions can also be easily simulated in the laboratory. Early scientific experiments with frogs in the late nineteenth and early twentieth centuries paved the way for cloning investigations in recent years.

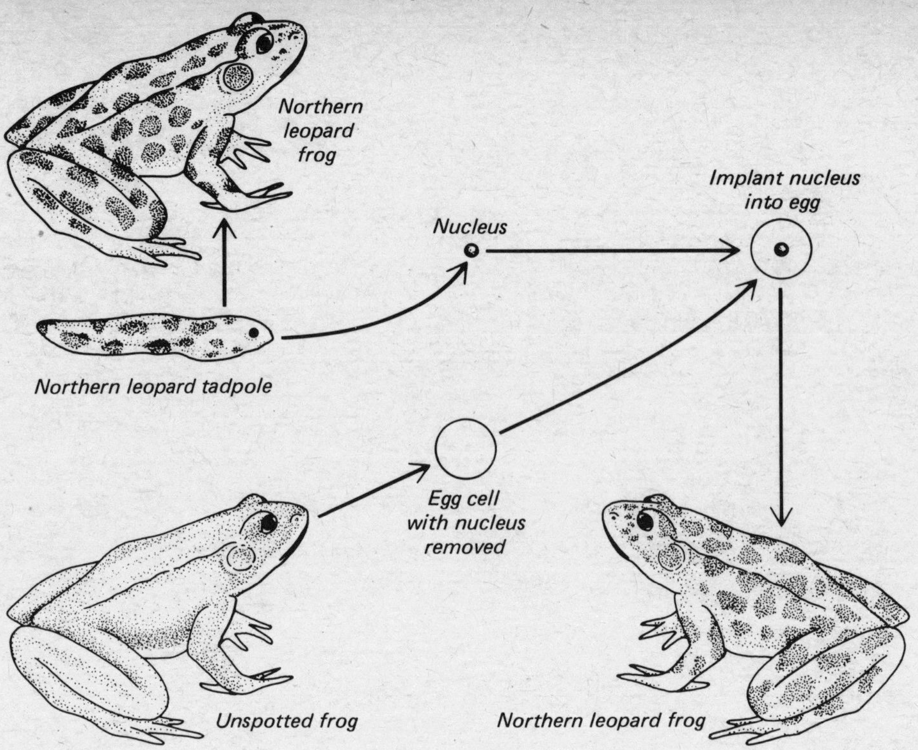

A classic example of frog cloning was performed by J. B. Gurdon of Oxford University. (Follow the steps in his procedure as shown in the illustration above.) First, the nucleus was removed from a tadpole's intestinal cell. Then he removed the nucleus from a frog's egg cell. (The frog and the tadpole were different species.) Next he implanted the nucleus from the tadpole's intestinal cell into the egg cell that lacked a nucleus. The implanted egg cell then developed into a frog *with the same traits as the tadpole whose nucleus was implanted in the egg*. Summarize this experiment by completing the Analysis section of your Data Sheet.

Part II. Investigation

Plant cloning is based on the fact that each plant cell has the potential of forming a new plant. In this laboratory exercise, you will propagate (increase the number of plants) by cloning a Boston fern plant. Remember that the purpose of plant cloning is to produce large numbers of genetically identical and superior plants. Be sure the Boston fern plant (one per class) is a healthy, well-branched plant.

The procedure that follows is specifically designed for a Boston fern plant. Each plant — even varieties within a species — has definite requirements for successful cloning.

NAME DATE

STAGE 1

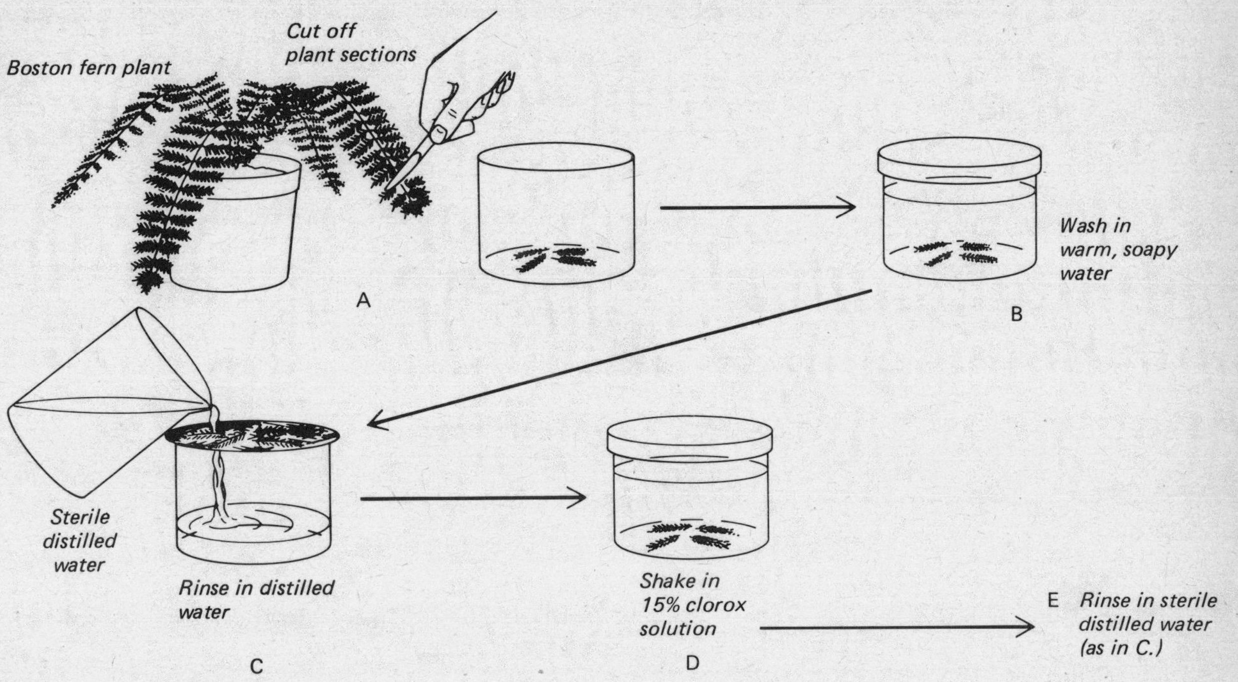

Preparation: Wash and sterilize scalpel, forceps, and glassware. During the sterilizing of the plant sections, dip the ends of forceps and scalpel in alcohol and then flame-sterilize them. CAUTION: ALCOHOL IS EXTREMELY FLAMMABLE.

Stage 1: Use a scalpel to cut off 2.5-cm stem and leaf sections of the Boston fern. If runners (small shoots that grow near the base of the plant) are present, cut off 2.5-cm end sections. (There should be 50 plant fragments per class for use in the 50 Nutri-Tubes.)

Wash the plant sections in warm, soapy water. Rinse with sterile distilled water as shown in the illustration above. Then place the plant sections in a 15% chlorine bleach solution. Shake the container for about 5 minutes. Then thoroughly rinse the fragments with sterile distilled water.

Use forceps to place a plant section in each Murashige Organics Medium Nutri-Tube. (This gelled medium contains mineral nutrients, sugars, and vitamins necessary for starting a plant culture.) Recap each tube and place the tray of tubes under fluorescent lights for 12 hours/day. *Room temperature must be maintained at all times.* Observe and record growth results for the next 2–3 weeks in your Data Sheet chart.

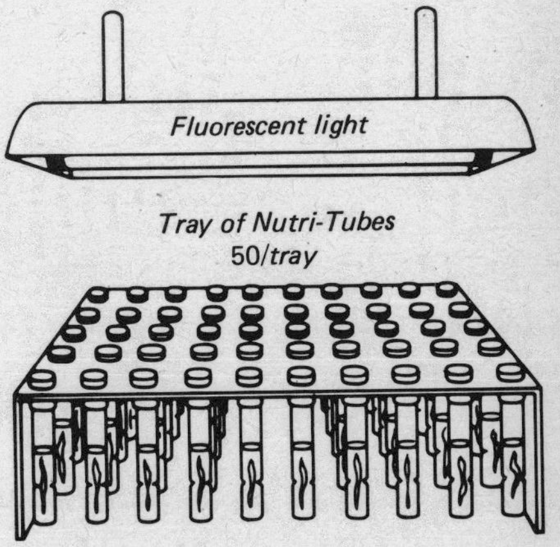

Plant culture setup

Stage 2: Use sterilized forceps to remove plant sections from Nutri-Tubes. Insert these sections into the Fern Multiplication Nutri-Tube Medium. (This culture medium contains chemical hormones called auxins and cytokinins, which stimulate root and bud growth.)

Place the tray of Nutri-Tubes under fluorescent lights and maintain room temperature. Observe and record results for a 3-week period on your Data Chart.

Stage 3: Sterilize forceps, scalpel, and glassware. Remove plant sections with forceps and place in Petri dishes. Divide plant sections in half and place in 100 tubes (2 trays) of Murashige Organics Medium. Place the two plant culture trays under fluorescent lights. *Note: Increase the intensity of the fluorescent lights by adding lights or using lights of higher wattage*. Record your results for the next 3 weeks on your Data Chart.

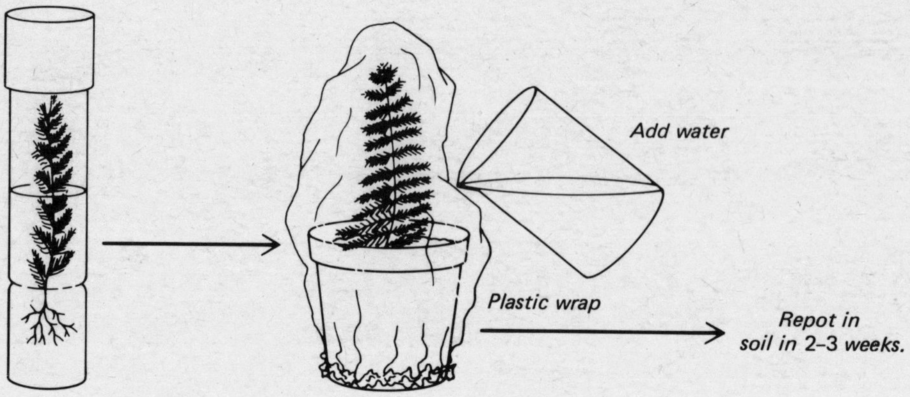

Remove new plant growth from tubes and pot them in vermiculite. Water and cover with plastic wrap. Place the new plants in bright, indirect sunlight (not in full sun). Gradually uncover plastic wrap as growth proceeds. Repot in soil in 2–3 weeks. Complete Data Sheet questions 1 through 4, and record the number of successful clone plants propagated by your class.

Investigations On Your Own

(1) Review the various ways plants can be propagated — division, air layering, stem and leaf cuttings, runners, and cloning of plant sections. Detail the advantages and disadvantages of cloning.

(2) Do a research report on the historical background of scientific cloning. Include a discussion of ethical questions concerning cloning of mammals.

NAME ______________________________ DATE ______________

Cloning 12

DATA SHEET

Analysis

Carefully read the background material on frog cloning and the experiment of J. B. Gurdon as detailed in the Analysis section of your Procedure page. Then complete the following fact sheet.

THE PROBLEM (in question form) ______________________________

REASONS FOR USING FROGS: ______________________________

____

____

EXPERIMENT DESCRIPTION: ______________________________

____

____

____

CONCLUSION: ______________________________

____

Investigation

(1) Why is so much care taken to sterilize equipment and plant sections? ______________

____

(2) What is the difference between the Stage 1 and Stage 2 media? ______________

____

(3) Which chemicals are responsible for stimulating bud and root growth? ______________

____

(4) List the physical requirements for plant cloning. ______________

____

____

Observable Growth

Stage	First Week	Second Week	Third Week
1			
2			
3			
Vermiculite Planting			
Soil Planting			

Summary

(1) What is a clone? ______________________________

____

(2) How does cloning differ from sexual reproduction? ______________________________

____

____

(3) Why are plants cloned? ______________________________

____

(4) Why are animals cloned? ______________________________

____

____

NAME DATE

Adaptation – Organic Variation 13

MATERIALS

INVESTIGATION▶

- 25 clear acetate discs
- 25 yellow acetate discs
- 25 blue acetate discs
- 25 red acetate discs
- various living and preserved specimens (plant and animal)

OBJECTIVES▶

- To demonstrate the meaning of the term adaptation
- To visualize survival of the fittest
- To understand how gene pools affect adaptability

Problem

What does adaptation mean?

An adaptation is an inherited variation that helps an organism to survive. Natural selection is the process by which the best adapted organisms survive. When the organism survives, it has an improved chance for reproducing others of its own kind. Charles Darwin developed the theory of adaptation as nature's way of selecting for the "survival of the fittest."

Procedure

Part I. Analysis

Study the drawings on your Data Sheet page of the light- and dark-colored peppered moths. Read the following background story of the peppered moths in England before answering questions 1–9.

Before 1845, most of the peppered moths in Manchester, England were light in color. They were a difficult prey for birds because they blended with the light-gray bark of the trees. The few black peppered moths stood out in their surroundings and were an easy target for their predators.

But in 1845, the Industrial Revolution created a different situation. Smoke and soot darkened the trees. The light-colored peppered moth population fell sharply and the dark-colored peppered moth population rose dramatically. Can the changes in moth population be explained by means of natural selection?

Part II. Investigation

(A) Use the entire floor area of the classroom as a habitat for a population of acetate "animals." These animals are all of the same species. The variation in the color of the discs representing the "animals" is due to different genetic inheritance. The color of each "animal" (disc) is homozygous for the color trait. The following are the genotypes for each "animal":

COLOR	GENOTYPE
clear	CC
red	rr
blue	bb
yellow	yy

Note that the clear "animal" is homozygous dominant while the other colors are recessive to clear.

Each individual in the class will act as a predator searching for food. The prey will be the acetate animals. You will be given 3 minutes to search the area for "food." Collect as many acetate animals as possible. When the time is up, reassemble and count the total number of acetate discs found by all members of the class. Fill in the chart on your Data Sheet page with your results.

Prepare a bar graph indicating the frequency of color genes remaining in the acetate animal population. Complete your investigation by answering questions 10–17.

(B) The best way to learn about adaptation is to examine organisms for the features that help to make them successful. Go to each of the observation areas and observe the living or preserved plants and animals.

In the live-animal area, observe the mouthparts, the skin, the means of locomotion, and the color of the organisms.

In the live-plant area, compare at least two different kinds of plants such as a cactus and a geranium. Note the differences in leaves and stems. In what ways are they adapted to different environments?

The preserved or dried specimens should be examined thoroughly to find plant and animal adaptations. Remember to think of the environments each of these specimens once inhabited. What features helped them to survive?

Report your observations by answering questions 18–25 on your Data Sheet.

Investigations On Your Own

(1) Organisms are adapted for living in their particular environment, but some seem to exhibit a greater adaptation than others. Prepare a report on two organisms — one a plant and the other an animal. Include in your discussion reasons why these organisms are considered to be so highly adapted. What changes would have to occur in their environment to make them less adapted and less able to compete successfully with other organisms?

(2) Do a research report on the naturalist Charles Darwin. Explain his discoveries during the voyage of the *Beagle*. What was the reaction of other scientists to Darwin's theory of natural selection? Include the reasons why Darwin's new insights are still considered remarkable.

NAME DATE

Adaptation – Organic Variation 13

DATA SHEET

Analysis

(1) Describe the moth population before 1845. ____________________

____

(2) What environmental conditions favored the light-colored peppered moths? ____________________

____

(3) What animal is the natural predator of peppered moths? ____________________

(4) Which peppered moths were easier prey before 1845? ____________________

____

(5) What happened during 1845 to change the environment? ____________________

____

____

(6) Which kind of peppered moth was able to survive the environmental changes?

____

(7) Which moths reproduced more of their own kind after 1845? ______

______

(8) Explain why the moths (your answer to 7) were so favored. ______

______

______

(9) How would Darwin explain the story of the light- and dark-colored moths? ______

______

______

______

______

Investigation

(A) Go on a "hunt" for acetate "animals" as described in the Procedure page.

(10) What genes are present in the gene pool of the acetate animals? ______

______

(11) What is the gene frequency of each color gene in the population of acetate animals?

______

Record your findings in the chart below.

Acetate Animals	Number Originally	Number Found	Number Left on Floor	Frequency of Genes Left
Clear				
Red				
Yellow				
Blue				

(12) Which "animals" were most difficult to find? Explain. ______

______

______

(13) Explain which acetate animals were the most fit. ______

______

NAME DATE

(14) Which acetate animals were the most poorly adapted? ____________

Why? ____________

____

(15) What will eventually happen to the genes of the most poorly adapted animals?

____

____

In the grid below, prepare a bar graph indicating the frequency of color genes remaining in the acetate animal population.

(16) As far as the predators (you) are concerned, which of you will live and which of you will die? ____________

____

(17) List several factors that make the predators better adapted for hunting. ____________

____

(B) Examine the living and preserved specimens at each of the observation areas.

Living Animals:

(18) What body organs help these animals to feed? ____________

____

____

(19) Describe the skin or outer covering of each of the animals. ______________________

____

(20) In what ways are these outer coverings adapted to the organisms' environments?

____

____

____

(21) Describe each animal's means of locomotion. ______________________

____

____

Living Plants:

(22) Compare the leaves of the cactus and the geranium. ______________________

____

(23) Compare their stems. ______________________

____

(24) Explain why each is adapted to its environment. ______________________

____

____

____

Preserved or Dried Specimens:

(25) Describe the features of these specimens that highlight the organism's adaptation to its environment. ______________________

____

____

Summary

(1) What does it mean for an organism to be adapted to its environment? ____________

____

____

(2) Explain the concept of the "survival of the fittest." ______________________

____

NAME DATE

Classifying Organisms 14

MATERIALS

ANALYSIS▶
- tracing paper, scissors, construction paper

INVESTIGATION▶
- freshwater fish specimen(s)
- classification key

OBJECTIVES▶
- To understand how to group similar objects and living things
- To learn how a classification key is constructed
- To use a classification key to identify unknown organisms

Problem

How are organisms classified?

Biologists group organisms according to structural similarities. By comparing characteristics of organisms, a pattern of classification can be understood. The classification groups in biology are: kingdom, phylum, class, order, family, genus, and species. The branch of science that deals with classifying organisms on the basis of their natural relationships is called taxonomy.

Procedure

Part I. Analysis

(A) Trace and cut out the figures drawn below from construction paper. Then follow the directions on the next page to classify these shapes into various categories.

NAME DATE

In this Analysis activity **(A)**, you will apply the classification categories used in taxonomy to the figures drawn on the first page of *Classifying Organisms*.

Assemble the figures into two groups based on a common characteristic. For instance, put all figures with curved lines into one group. The second group will then be of figures with only straight lines.

You should now have in the straight-line group 12 straight-line figures: one rectangle and one triangle with lines projecting from them and 10 others that are shaded or unshaded triangles, squares, or rectangles. Your second major group of figures should be those with curved lines — representing the other phylum.

Using the characteristic of lines projecting from the figures, divide the straight-line figures and the curved figures into two smaller groups.

Then separate the shaded figures in each large group from the unshaded figures. Separate the triangles from the squares and rectangles; separate the circles from the half- or double-circles. Finally, separate the figures on the basis of size.

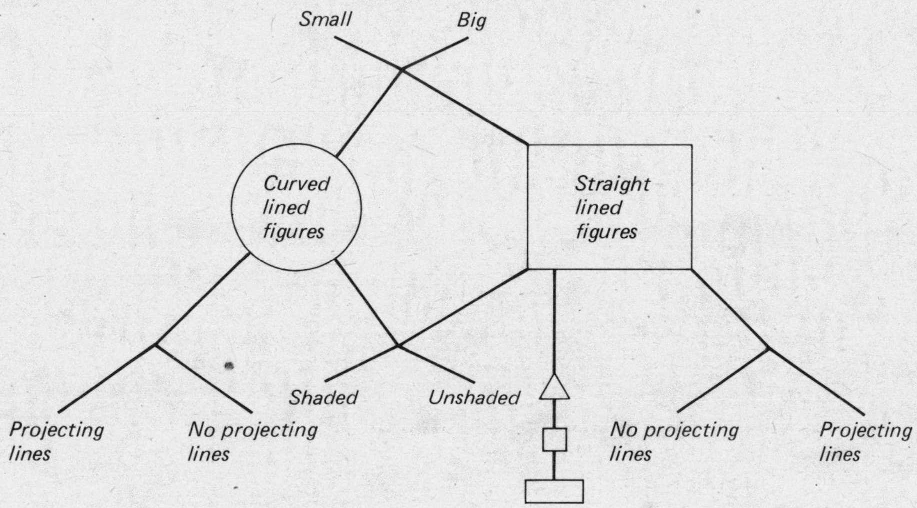

Classifying figures

On your Data Sheet, answer questions 1–4 and fill in the chart with the number of each figure for each classification group.

(B) Examine the drawings of plants and animals. Separate the organisms into two large groups. Then continue to divide them into smaller and smaller groups that are related in a specific way. Refer to your text as you categorize the different groupings.

Complete your classification system by answering questions 5 and 6 on your Data Sheet.

Part II. Investigation

Closely examine one of the drawings of native fish on page 90. Read both statements listed under number 1 in the classification key. One of these statements should describe the fish you have chosen; the other should not. Refer to the number after the statement that fits your fish and look for that number in the key. Again select the statement that describes the fish that you picked. Continue through the key until you come to a name after one statement. This should be the name of the fish you picked. Practice using the key to identify several of the fish shown.

Example:

Suppose you want to find the name of fish number 2. Look at the classification key. Note that each numbered item presents two possibilities. You see that the fish has no scales, or at least you cannot see any scales. So you should choose item 1b. This refers to number 12. So you go down the page to number 12. The fish is not elongated or snakelike (item 12b), so you go to number 13 of the key. The fish you are classifying has barbels (see below: "Terms Referring to the Structure of Fish") growing from its lip and the top of its head (item 13a), so you go to number 14 of the key. Since the fish has a caudal fin that is rounded, and a blunt head, you see that it is the *Bullhead catfish* (also known as *horn pout* in some parts of the country).

After you have identified the 18 fish shown in the picture, list their names on the Data Sheet, questions 7–24.

Now determine the identity of an unknown freshwater fish specimen. Use the key to help you in your search. Answer question 25.

Terms Referring To The Structure Of Fish

barbel — a fleshy projection from the lips or head
scales — overlapping outgrowths of the skin

Fins

adipose — a small fin on the top mid-line of the body near the tail fin
anal — a fin along the lower mid-line of the body near the tail fin
caudal — tail fin
dorsal — the fin or fins along the top mid-line of the body
pectoral — the paired fins nearest the head, corresponding to front legs or arms
pelvic — the paired fins nearest the tail, corresponding to hind legs

Investigations On Your Own

Select commonly seen groups of related objects (automobiles, canned goods, etc.) and classify them into the major categories. Construct a key to their identification. Try your classification key with others to see how well it works.

NAME ________________________________ DATE ________________

Classifying Organisms 14

DATA SHEET

Analysis

(A) Follow the Procedure page directions for grouping the various figures that you traced.

(1) List some of the characteristics of the figures. ________________________________

____

____

(2) If considered in biological terms, what classification category would each individual figure represent? ________________________________

(3) What classification category would the entire group of figures represent? ________

____

Fill in the number of figures for each category in the chart below.

	Phylum	Class	Order	Family	Genus	Species
		(no projecting lines)	(unshaded)	(triangles or circles)	(squares or half-circles	(small squares or half-circles)
Straight-lined Figures						
Curved-lined Figures						

(4) Explain how you could further divide the figures into different classes, orders, families, etc. ________________________________

____

(B) Separate the drawings of organisms on page 86.

(5) What are the two major groupings or phyla? ________________________________

(6) Explain how you separated each phylum. ________________________________

____

____

____

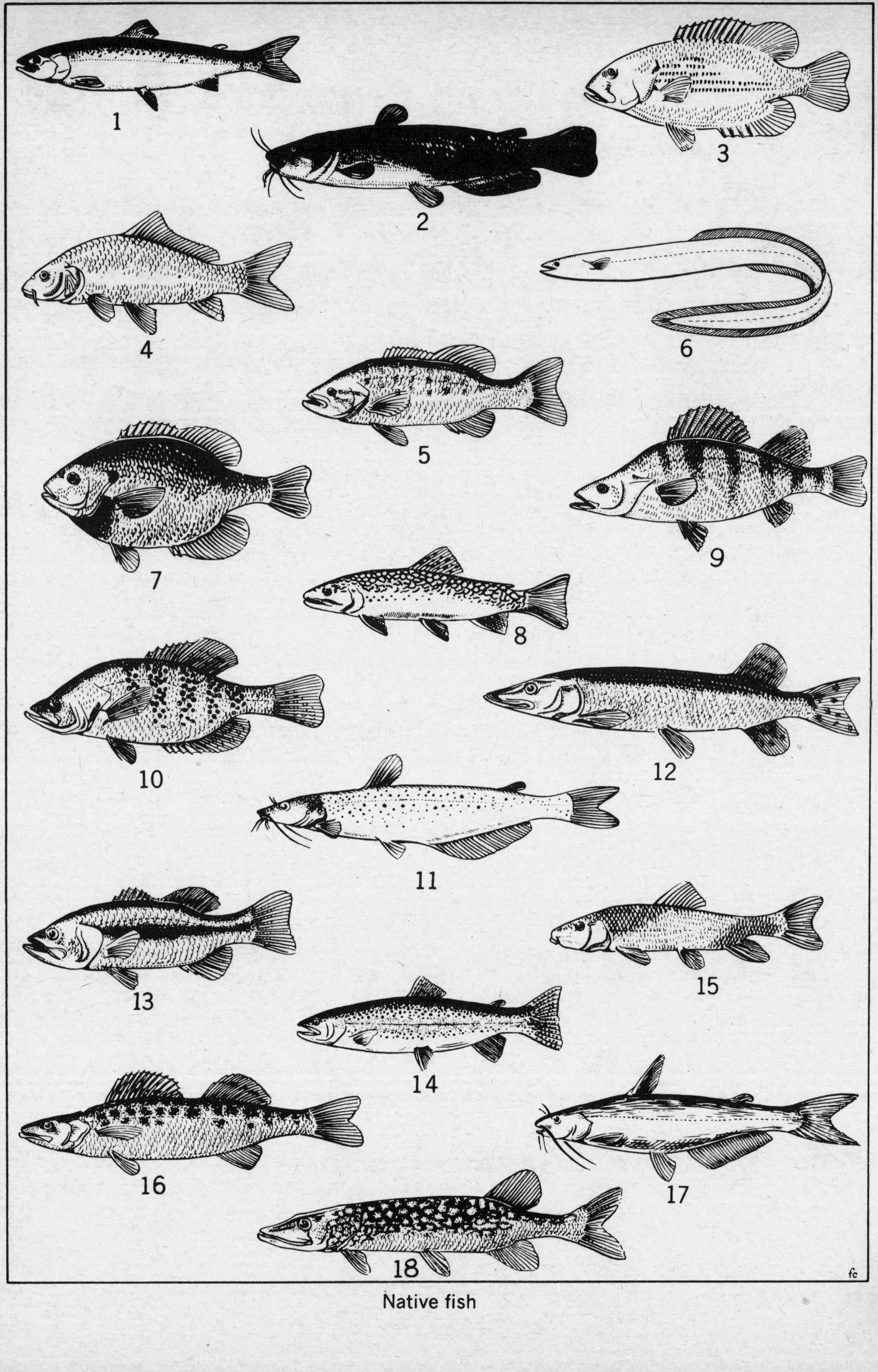

Native fish

NAME DATE

CLASSIFICATION KEY TO CERTAIN FISH

1a.	Body noticeably covered with scales	2
1b.	Scales not covering body or too small to be seen	12
2a.	Dorsal fin single	3
2b.	Dorsal fins two or more, joined or separated	6
3a.	Body more than four times as long as broad (top to bottom); front edge of dorsal fin far back on body; mouth large, hinge back of eye	4
3b.	Body less than four times as long as broad; front edge of dorsal fin about midway between head and tail; mouth not large, hinge in front of eye	5
4a.	Dark lines forming netted design on body; fins not spotted	*Pickerel*
4b.	Body covered with yellow spots; fins spotted	*Northern pike*
5a.	Mouth turned downward; barbels absent; dorsal fin not elongated	*White sucker*
5b.	Mouth not turned downward; barbels present; dorsal fin elongated	*Carp*
6a.	Two dorsal fins separated, the anterior spiny and the posterior soft	7
6b.	Two dorsal fins united, forming an anterior spiny portion and a posterior soft portion	8
7a.	Top of head concave, forming a hump in front of dorsal fin; dark vertical bars on body	*Yellow perch*
7b.	Top of head not concave, body sloping to dorsal fin and not forming a hump; dark blotches on body	*Wall-eyed pike*
8a.	Body more than three times as long as broad	9
8b.	Body less than three times as long as broad	10
9a.	Hinge of jaws behind the eye; notch between spiny and soft dorsal fin deep and nearly separating into two fins	*Large-mouth black bass*
9b.	Hinge of jaws below the eye; notch between spiny and soft dorsal fin not nearly separating into two fins	*Small-mouth black bass*
10a.	Mouth large, hinge below or behind eye	11
10b.	Mouth small, hinge in front of eye	*Bluegill*
11a.	Five to seven spines in dorsal fin; dark spots forming broad vertical bars on sides	*White crappie*
11b.	Ten or more spines in dorsal fin; sides flecked with dark spots	*Rock bass (Redeye)*
12a.	Body much elongated and snakelike; dorsal, caudal, and anal fins continuous	*Eel*
12b.	Body not elongated and snakelike; dorsal, caudal, and anal fins separate; adipose fin present	13
13a.	Barbels growing from lips and top of head; head large and broad	14
13b.	Barbels lacking; head not large and broad	16
14a.	Caudal fin deeply forked; head tapering	15
14b.	Caudal fin rounded or slightly indented but not forked; head blunt	*Bullhead catfish*
15a.	Dorsal fin rounded at top; body silvery, speckled with black markings	*Channel catfish*
15b.	Dorsal fin long and pointed at top; body bluish-gray without speckles	*Blue catfish*
16a.	Caudal fin deeply forked; back not mottled and with few spots	*Atlantic salmon*
16b.	Caudal fin square or slightly indented; back mottled or spotted	17
17a.	Back and caudal fin spotted; broad horizontal band along sides	*Rainbow trout*
17b.	Back mottled with dark lines; caudal fin not spotted; fins edged with white	*Brook trout*

Investigation

Identify the 18 fish according to the classification key and the directions on your Procedure page. Fill in the name of each fish in the spaces below.

7. ______________________ **16.** ______________________

8. ______________________ **17.** ______________________

9. ______________________ **18.** ______________________

10. ______________________ **19.** ______________________

11. ______________________ **20.** ______________________

12. ______________________ **21.** ______________________

13. ______________________ **22.** ______________________

14. ______________________ **23.** ______________________

15. ______________________ **24.** ______________________

(25) Name the unknown fish that you identified and explain why you believe your identification is accurate. ______________________

____

____

Summary

(1) Based on what you have learned in this Investigation, discuss how classification is a useful tool for a biologist. ______________________

____

____

____

Fill in the blanks:

(2) A group of closely related species is a ______________________.

(3) The largest of the classification categories is the ______________________.

(4) The most specific of the classification groups is the ______________________.

(5) A group of closely related classes is a ______________________.

(6) The subdivision of an order is the ______________________.

(7) A ______________________ is composed of several closely related orders.

NAME DATE

Bacterial Viruses 15

MATERIALS

INVESTIGATION▶

- glass marking pencil
- bacterial loop
- test tubes
- stock culture of *Streptococcus faecalis*
- stock culture of *Streptococcus faecalis* phage
- sterile cotton plugs
- test tube rack
- Bunsen burner
- S F broth
- agar
- sterile Petri dishes
- distilled water
- incubator set at 37°C
- 250 milliliter beaker
- sterile cotton swabs

OBJECTIVES▶

- To determine the structure of a phage virus
- To understand the life cycle of a virulent phage
- To learn some bacteriological techniques
- To observe the effects of a phage

Problem

What is the structure and composition of the bacteriophage virus and how does it destroy a bacterium?

Bacterial viruses invade only certain kinds of bacteria. The electron microscope has the ability to magnify 100,000 times or more. It has been a valuable tool that has enabled biologists to determine the structure of bacteriophages.

During the lytic cycle, phages destroy bacteria in six basic steps: *normal bacterium, phage attaches, phage DNA forced through cell wall, phage DNA reorganizes bacterial DNA and protein synthesis builds more phage DNA, bacterium builds more virus particles, cell wall breaks liberating phages*.

Procedure

Part I. Analysis

Turn to the Data Sheet. Explain the lytic cycle shown in the diagram.

Part II. Investigation

(A) Growing bacteria and phages in broth.

In this Investigation you will observe the results of *Streptococcus faecalis* phage on the bacteria grown in a broth. *Streptococcus faecalis* is a bacterium found in the human intestine and may be found in polluted water. The broth that you will be using contains an indicator, Brom-Cresol Purple, and dextrose. As the bacteria grow, lactic acid is produced, and the purple indicator changes to a yellow color. If a bacteriophage lyses the bacteria, the acid is not produced, and the broth remains purple.

Take two test tubes containing sterile S F broth. Using a glass marking pencil, label one "C" for *control* and one "E" for *experimental*. Look at the diagrams below and, before continuing, be sure you understand the procedure you will use.

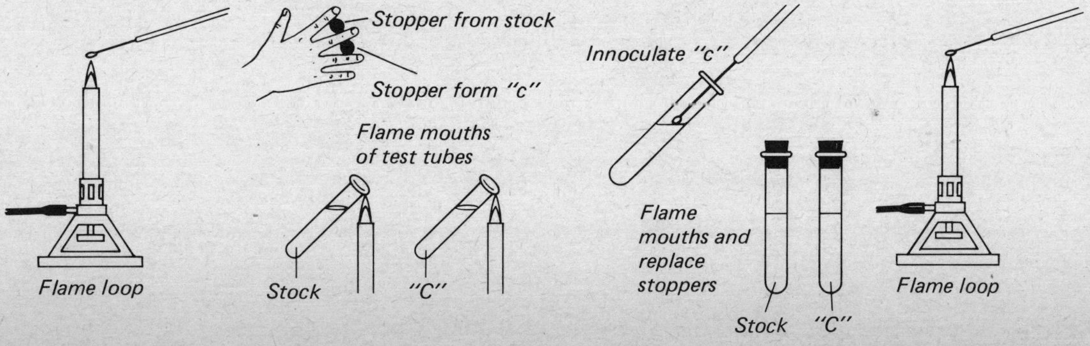

1. Pass the bacterial loop through the flame until it glows red.
2. When cool, transfer a loopful of the *Streptococcus faecalis* bacteria from the stock solution to the control tube (C).

3. Flame the loop, cool, and transfer a loopful of the *S. faecalis* from the stock solution to the experimental tube (E).
4. Flame the loop, cool, and transfer a loopful of the *S. faecalis* phage from the stock solution to the experimental tube (E).
5. Wash your hands thoroughly.
6. Place both the control and experimental tubes in a 37°C incubator. Examine the tubes at 24 hours and again at 48 hours. Record and interpret your results on the Data Sheet by answering questions 1–5.

(B) Growing bacteria and phage on agar.

Use a commercially prepared agar to prepare medium. Add 1.5 grams dehydrated S F broth to each 100 grams commercially prepared agar, and mix well. Prepare enough medium for a 1-liter flask. Stir the mixture and put the flask on a burner. Continue to stir *constantly* until the solution comes to a boil.

Sterilize the prepared agar medium in an autoclave or pressure cooker. Remove the flask of medium from the sterilizer, and take out the cotton plug. Flame the mouth of the flask.

Using sterile plastic Petri dishes, pour them as directed. Raise the cover slightly and add enough sterile medium to cover the bottom of the dish to a depth of about one-half centimeter. Replace the cover as quickly as possible. When all dishes have been poured, put them on a level surface and allow them to gel.

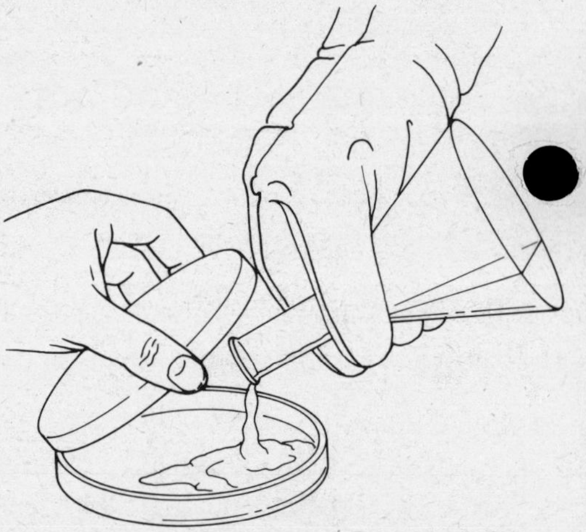

When cooled, take two plates and treat them as follows:
1. Using a wax glass marking pencil, make a "C" on one plate for control and an "E" on another plate for experimental.
2. Take the stopper from the stock solution of *S. faecalis*, flame the mouth, and dip a sterile cotton swab into the culture. Flame the mouth and return the stopper.
3. Gently swab over the agar as shown in the diagram. Remember to raise the cover only slightly as in the diagram above showing the pouring of the plates. Put the swab into the 250 milliliter beaker marked "disposal," to be sterilized.

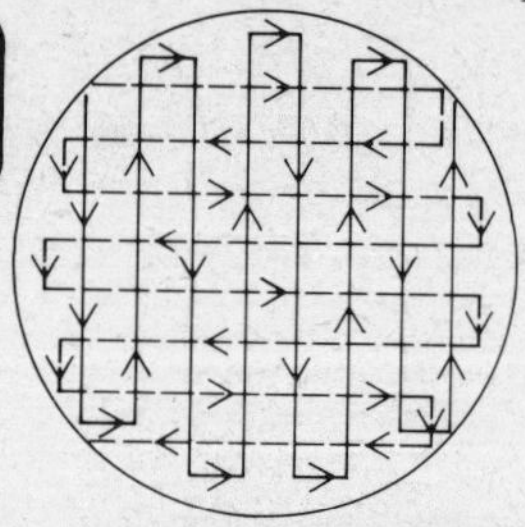

Inoculating petri dish

4. Repeat the procedure by inoculating the experimental Petri dish.
5. Now, flame a bacterial loop. When cooled, dip it into the stock solution of the *S. faecalis* phage culture. Raise the top of the experimental dish and touch the loop to the center of the agar only. Replace the cover quickly and flame the loop again.
6. Incubate both Petri dishes at 37°C. *Wash your hands thoroughly.*
7. Examine the Petri dishes at 24 hours and again at 48 hours. Record your results and conclusions by noting color changes and answering questions 6–10 on the Data Sheet. Then, place the plates where directed for disposal.

NAME ______________________ DATE ______________

Bacterial Viruses 15

DATA SHEET

Analysis

LYTIC CYCLE

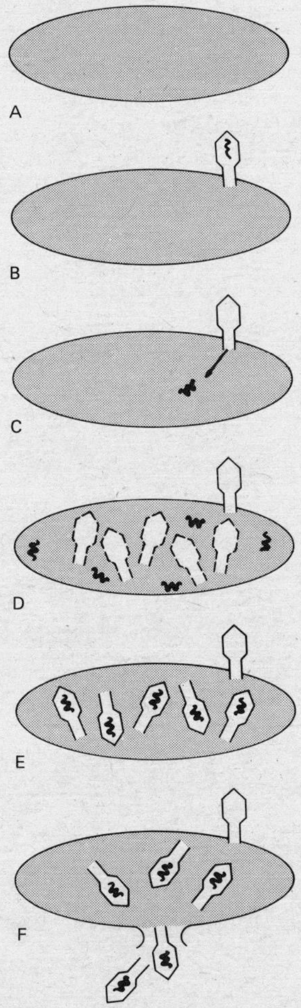

Step **(A)** ______________________

Step **(B)** ______________________

Step **(C)** ______________________

Step **(D)** ______________________

Step **(E)** ______________________

Step **(F)** ______________________

Investigation

A. Growing bacteria and phage in broth.

(1) What color is tube "C"? ______________________

(2) What caused this color? ______________________

(3) What color is tube "E"? ______________________

(4) What caused this color? ______________________________

____

(5) From your understanding of the lytic cycle of a phage, explain what happened in the experimental tube. ______________________________

____

____

____

B. Growing bacteria and phage on agar.

(6) What is the color of both Petri dishes at the beginning of the experiment? ________

____

(7) Using colored pencils or merely labels, indicate any color changes in the diagrams below.

(8) Explain your results. ______________________________

____

____

____

____

(9) From this Investigation, did you see any evidence that bacterial cells were destroyed? Explain your answer. ______________________________

____

____

____

____

(10) Explain the *control* and the *experimental* in the Investigation with the broth.

____

____

____

NAME DATE

Identification of Bacteria 16

MATERIALS

INVESTIGATION▶

- prepared slides of the various forms of bacteria
- distilled water
- microscope
- 15–20 dried beans
- 400-milliliter beaker
- depression slides
- cover glasses
- petroleum jelly
- Bunsen burner
- inoculating loop
- toothpicks

OBJECTIVES▶

- To identify three forms of bacteria
- To observe the motility of bacteria in a bean infusion

Problem

How can bacteria be identified and studied?

Microbiologists classify bacteria into three groups: *coccus*, *bacillus*, and *spirillum*. The characteristics of these bacteria can be easily identifiable under the microscope.

Living bacteria appear to move because the bacteria are moved with the molecules in a medium. This movement or motion is referred to as *Brownian movement*. Flagellated bacteria have true movement, or motility.

Procedure

Part I. Analysis

Turn to the Data Sheet. Refer to your text to help you identify the three basic forms of bacteria. Answer questions 1–10 on the Data Sheet.

Part II. Investigation

(A) Obtain and examine the prepared slides of the various forms of bacteria under both low and high power of the microscope. Draw each form of bacteria that you observe, in the space provided on the Data Sheet. Label your drawings. Answer questions 11–15 on the Data Sheet.

(B) Motility can be seen by preparing a bean infusion and examining the bacteria in this infusion. To prepare a bean infusion, place 15–20 beans in half a beaker of water and boil for a few minutes. Add water to fill the beaker. Set the infusion in a warm place, exposed to the air, for 2 or 3 days.

Using a toothpick, place a thin film of petroleum jelly around the edge of a depression of a culture slide. Then use an inoculating loop to transfer a loopful of bean infusion to the center of a cover glass. Hold the culture slide, hollowed side facing down, over the drop. Press the slide gently against the cover glass so that the petroleum jelly makes contact. Gently invert the slide against the cover glass so that the petroleum jelly makes contact. Turn the slide over and examine with the high power of the microscope. (This procedure is shown on the following page.)

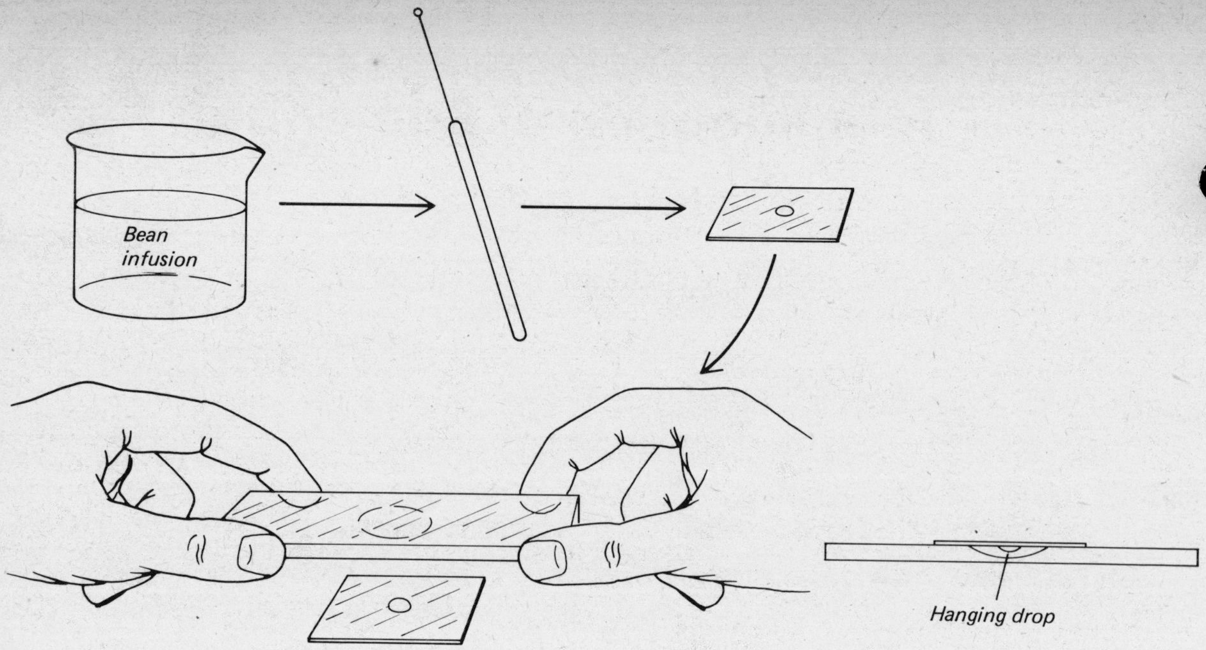

The infusion should be filled with bacteria. The organisms will be more visible if you darken the field by closing the diaphragm. Examine the preparation and answer the questions on the Data Sheet. Look for both motility and Brownian movement. See how many different forms of bacteria you can identify. Answer questions 16–20 on the Data Sheet.

Investigations On Your Own

(1) In the above Investigation, you examined bacteria in a bean infusion. You may be interested in making infusions with other substances such as hay, grass, or rice.
(2) There are certain *special stains* that are used by microbiologists to identify particular structures or types of bacteria. You may wish to investigate these stains and design an experiment that would test one or more of these stains.

NAME _______________________________ DATE _______________

Identification of Bacteria 16

DATA SHEET

Analysis

Identify each of the three basic forms of bacteria in the spaces provided. Then describe any special characteristics you observe in each form.

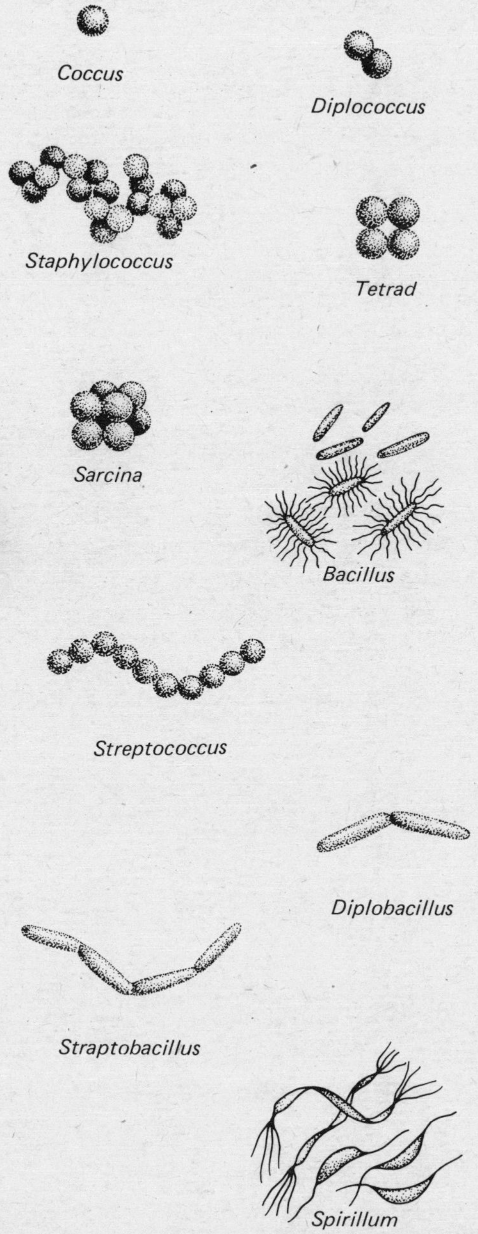

BASIC FORM *SPECIAL CHARACTERISTICS*

(1) ______________________________

(2) ______________________________

(3) ______________________________

(4) ______________________________

(5) ______________________________

(6) ______________________________

(7) ______________________________

(8) ______________________________

(9) ______________________________

(10) ______________________________

Investigation

In the spaces provided, draw each form of bacteria that you have observed in the prepared slides of various bacteria. Label your drawings.

(11) Should you view the bacteria under low or high power?

(12) Describe each of the three forms of bacteria.

(13) Which forms of bacteria are more common?

(14) Do you find colonies of bacteria?

(15) What kind of bacteria make up the colonies?

(16) Are motile (free-swimming) bacteria present in the infusion?

(17) What type of motion is demonstrated in your culture?

(18) Do the cells move individually or in groups?

(19) What structures enable many bacteria to move in the fluid?

(20) Can you distinguish any flagella on the bacteria?

Summary

(1) What characteristic is used to classify bacteria into three groups?

(2) Name the three groups of bacteria.

(3) What characteristics are used to classify each bacterial group?

NAME DATE

Antiseptics and Antibiotics 17

MATERIALS

INVESTIGATION▶

- about 3 common antiseptics, different ones
- 5% phenol solution
- filter paper
- forceps
- Petri dishes
- 3 Erlenmeyer flasks (250 mL)
- 500 mL Erlenmeyer flask
- nutrient agar
- nutrient broth
- cotton plugs for flasks
- sterile 1-mL pipettes
- bulb to fit 1-mL pipette
- paper punch
- sterile distilled water
- metric scale
- stock cultures of *Escherichia coli, Bacillus subtilis,* and *Proteus vulgaris*
- waxed pencil or gummed labels
- 6 media bottles
- cotton
- inoculating loop
- pressure cooker or autoclave
- dry-air oven
- Sensitivity discs — penicillin, streptomycin, terramycin, chloromycetin, aureomycin, and tetracycline

OBJECTIVES▶

- To determine the effect of antiseptics on bacteria
- To determine the effect of various antibiotics on the growth of bacteria
- To determine sensitivity and resistance of bacteria to antibiotics

Problem

What is the effect of antiseptics and antibiotics on bacteria?
Antiseptics and antibiotics are chemicals that inhibit (stop or slow down) the growth of bacteria. Different antiseptics and antibiotics affect bacteria to varying degrees. Some of these chemical substances are only effective against certain bacteria.

Procedure

Part I. Analysis

In 1929, Sir Alexander Fleming made a surprising discovery. While culturing bacterial growths, he found that a mold had "contaminated" one section of a culture. But in the mold area, the bacteria *ceased* to grow! Years later, the mold — *Penicillium notatum* — was mass-produced as an antibiotic effective against certain bacterial infections.

Using your knowledge of the scientific method, describe an experiment you would perform to validate Fleming's "accident." Include in your report the following steps in the research method:

1. Define the problem — usually in the form of a question.
2. Gather information.
3. Form a hypothesis — a trial answer to the problem.
4. Experiment to text the hypothesis. (Include a control experiment that has all the factors *except* the one being tested.)
5. Observe and record data.
6. Draw conclusions.

Part II. Investigation

(A) In this section, you will determine the bactericidal action of various antiseptics. You will compare the effectiveness of each antiseptic with that of a 5% phenol solution, which is used as a standard.

Obtain a variety of antiseptics, including 5% phenol solution. Mark the bottoms of three Petri dishes, dividing each into quarters. Number the quarters 1 through 4. Mark one dish *A*, one *B*, and one *C*. Using a sterile 1-mL dropper, transfer 1 mL of the *Bacillus subtilis* inoculum to sterile Petri dish *A*.

Using another sterile dropper, prepare sterile Petri dish *B* with the *Escherichia coli* inoculum. Prepare sterile Petri dish *C* with *Proteus vulgaris* inoculum. Cool liquefied nutrient agar to 45°C and pour all of the dishes. Use care to diffuse the inoculum through the warm nutrient agar before it gels.

Using a notebook paper punch, punch small discs of filter paper (unless prepared discs are available). With forceps, dip 3 filter paper discs into the 5% phenol antiseptic. Drain excess liquid from the disc: do not drip antiseptic on the surface of the poured plate. Place one disc in the number 1 quadrant of each culture dish. Discs should be placed 2 cm from the rim of the Petri dish. Prepare 3 discs for each antiseptic. Place one disc of each antiseptic in one quadrant of each culture dish. Incubate all plates for 48 hours at 35°–37°C.

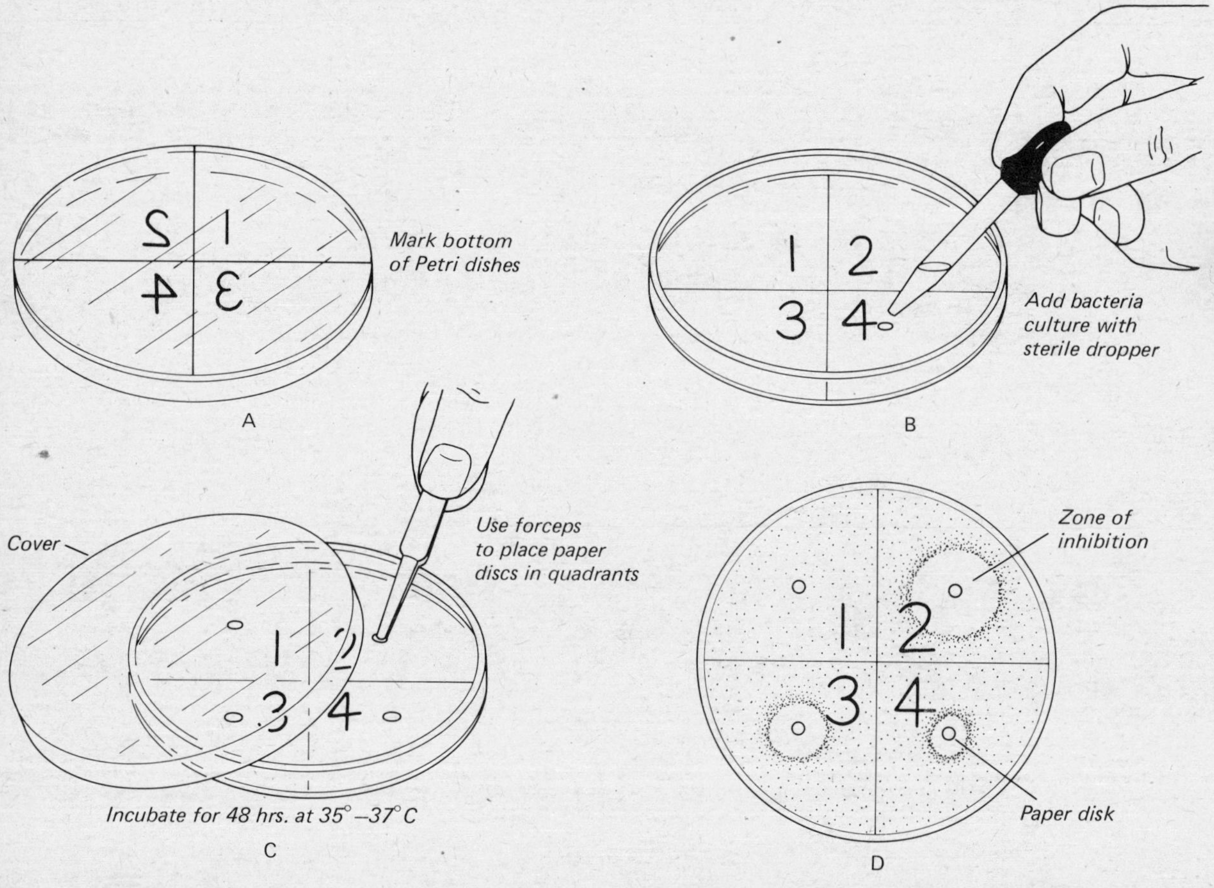

After incubation, examine the plates for evidence of a zone of inhibition around each disc. The zones appear as clear areas and indicate destruction of bacteria or inhibition of growth. Invert the dish and lay a metric ruler across each disc. Measure the width of the zone of inhibition on each side of the disc. Record your results in the Data Sheet Table and answer questions 7–9.

(B) In this section, you will use the disc technique to determine whether bacteria are sensitive or resistant to an antibiotic.

Identify each of the sterile Petri dishes with the name of the bacteria you are testing. Using a pipette bulb (not your mouth), transfer 1 mL of the inoculum to the appropriate Petri dish. Then pour 20 mL of the sterile nutrient agar that has been previously liquefied and cooled to 45°C into each Petri dish. Agitate the plates to diffuse the inoculum. Allow it to gel. (Let the agar gel before putting discs in the plates. Plates are not to be inverted during incubation since the discs might fall off the plates.)

After the agar has gelled, use sterile forceps and place the antibiotic discs (usually three levels of concentration are available) 2 cm from the edge of the dish. Place them at 12, 3, and 9 o'clock on the agar, as shown below. Test all of the antibiotics against all of the bacteria you have available. This will give you some interesting results for comparison of sensitivity. As a control, place a filter paper disc, saturated in the sterile distilled water, in the center of each plate. Invert the plates and incubate at 37°C for 48 hours.

After 48 hours, examine the plates and look for a zone of inhibition. This is a clear area formed around the disc and is due to the inhibitory action of the antibiotic.

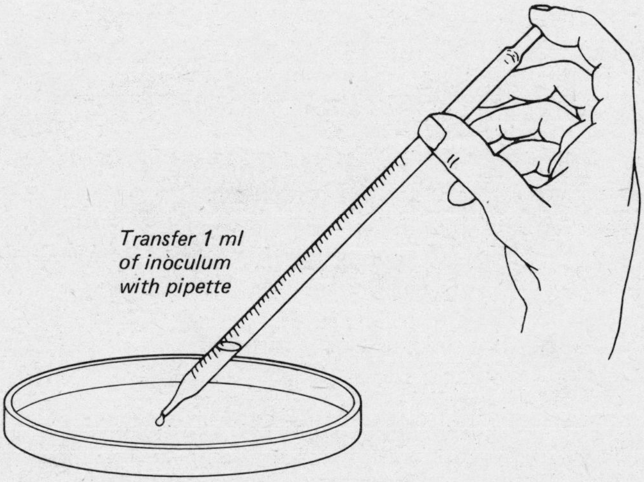

A

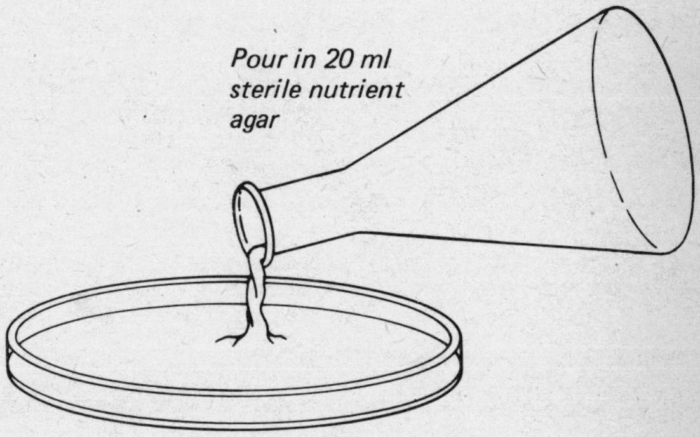

B

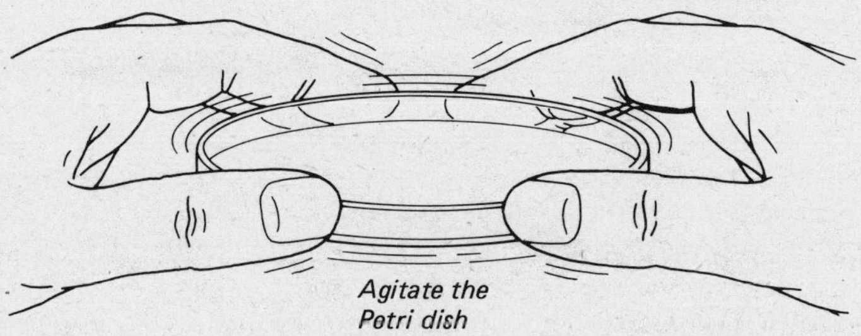

C

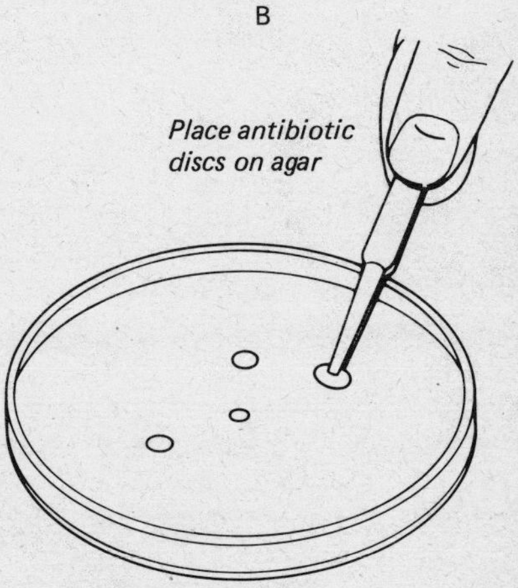

Invert plates. Incubate for 48 hrs. at 37°C.

D

Some of the variables in Investigation **(B)** are: the amount of the inoculum, the type of organism, the diffusion rate of the antibiotic, and the rate of deterioration of the antibiotic. Therefore, sensitivity is determined by the presence of a zone around any of the discs and not by the size of the zone. Consider the organism resistant if you find no zone on the plate. The organism is very sensitive to the antibiotic if there is a zone around the lowest level of concentration. An organism showing a zone around the disc of medium concentration, but not around that of the lowest, would be classified as sensitive or moderately sensitive. A zone around the disc of highest concentration indicates that the organism is only slightly sensitive to the antibiotic.

Now examine the plates and record the results as *resistant, slightly sensitive, sensitive,* or *very sensitive* in the appropriate column in the Table provided in the Data Sheet.

Investigations On Your Own

(1) Design an investigation testing various brand-name antiseptics to see which is the most effective in the inhibition of bacterial growth. Be sure to use an antiseptic substance as a control.

(2) Why are some bacteria more resistant to certain antibiotics than other bacteria? Attempt to answer this question through research and experimentation. At the beginning of the investigation, be sure to determine which bacteria are most resistant to a particular antibiotic. Then, attempt to find out why this is so.

NAME DATE

Antiseptics and Antibiotics 17

DATA SHEET

Analysis

Read the directions on your Procedure page before writing the following report.

(1) The Problem: ______

(2) Information I've Gathered: ______

______

(3) My Hypothesis: ______

______

(4) Description of My Experiment: ______

______

______

(5) Observations and Data: ______

______

______

(6) Conclusions: ______

______

Investigation

(A) Record the measurements of the inhibition zones of the different antiseptics.

	Antiseptic	Inhibition Zone (diameter)		
		Bacillus subtilis (A)	*Escherichia coli* (B)	*Proteus vulgaris* (C)
1.	5% phenol			
2.				
3.				
4.				

(7) Were the antiseptics that you used more or less effective than the control of 5% phenol solution? ______

______

(8) Were any of the bacterial cultures greatly inhibited by a particular antiseptic? If so, which ones? ______

______

(9) Were any of the bacterial cultures *not* affected by the antiseptics used? If so, which ones? ______

______

(B) Examine the effect of the various antibiotics on the different bacterial cultures. Record your results in the Table below. Identify the ones that were *resistant, slightly sensitive, sensitive*, or *very sensitive*.

	Bacillus subtilis **(Gram-positive)**	*Escherichia coli* **(Gram-negative)**	*Proteus vulgaris* **(Gram-negative)**
Penicillin			
Streptomycin			
Terramycin			
Chloromycetin			
Aureomycin			
Tetracycline			

Summary

(1) What does a clear zone around a disc indicate in both parts of this Investigation?

______

(2) Why is it important that aseptic techniques be used in both parts of this Investigation? ______

(3) What variable factors could affect the zones of inhibition in Investigation **(B)**?

______

______

(4) Why are antibiotics not equally effective against each type of bacteria? ______

______

NAME DATE

18 Microscopic Life

MATERIALS

INVESTIGATION▶

- pond water
- slides
- cover slips
- microscope
- methyl cellulose solution
- medicine droppers
- paramecia culture
- carmine red solution
- acetic acid
- dark paper
- matches
- spun glass
- pieces of cardboard
- salt crystals
- vinegar
- ink
- test tubes
- stoppers for test tubes

OBJECTIVES▶

- To identify the structures of the ameba, *Euglena*, and paramecium
- To observe the specialization of the paramecium
- To examine the response of paramecia to various stimuli

Problem

What kinds of microscopic life are found in pond water? Does a protist organism (such as the paramecium) respond to stimuli?

Pond water is teeming with all kinds of unicellular organisms called protists. Many of these organisms belong in the phylum *Protozoa* or "first animals." Some protozoans are clearly animallike, while others are more plantlike. Yet each of these one-celled organisms moves, takes in food, excretes waste products, reproduces, and performs other life functions.

Although paramecia do not have any specialized sense organs, they do respond to stimuli. They move away from extreme heat or cold, chemicals, or lack of oxygen; they move toward acid areas where bacteria, a food source, are often located.

Procedure

Part I. Analysis

Examine the outline drawings of ameba and *Euglena* on your Data Sheet page. Refer to your text in order to label the specialized features of each organism. Then answer questions 1–2 comparing these two common protists.

Part II. Investigation

(A) In this section, you will take a close look at the organisms that live in a drop of pond water. These organisms make up a community. Such a community is composed of many different types of organisms including algae, protozoans, and tiny invertebrates.

As you do this activity, refer to the illustrations in your text and the drawings to help you identify the organisms in your pond water sample. Prepare a number of wet mounts of pond water. Be sure to prepare a slide of the muddy water as well as the clear water.

Observe the slides first under the low-power objective of your microscope and then under the high-power objective. Compare the organisms on your slide with those in your text and in the drawings.

Identify and draw the organisms in the space provided on your Data Sheet.

In order to observe the protists more closely, prepare another wet mount and add a drop of methyl cellulose solution. Examine your slide under both low and high power. Try to distinguish the pond-culture producers from the pond-culture consumers. Complete your pond-water study by answering questions 3–8.

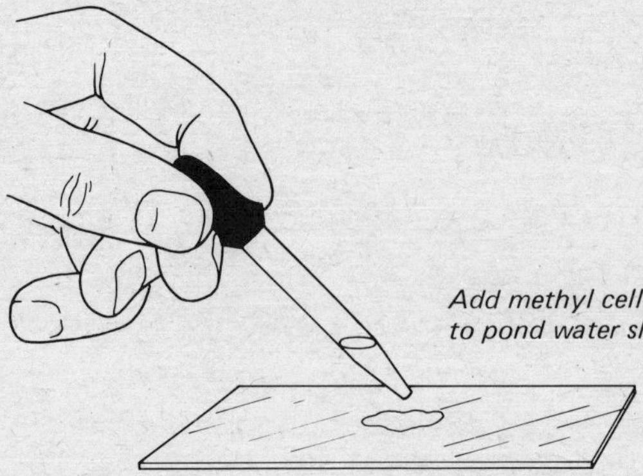

A

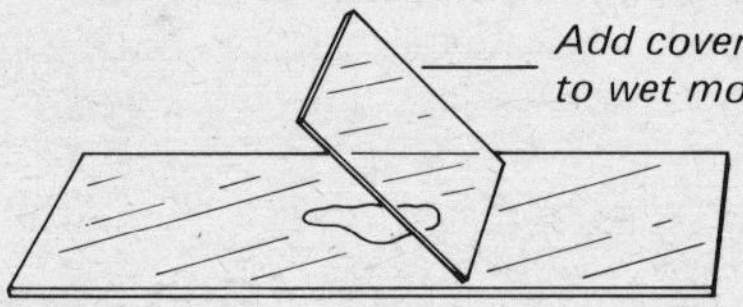

B

(B) Now you will examine a common protozoan, the paramecium. The paramecium is a unique organism with specialized structures for movement. It may appear to move more rapidly than it really does because of the magnification of the microscope.

Place a drop of scum from the paramecium culture on a slide. Cover with a cover slip. Examine the movement of the paramecia under the low power of the microscope. In the space provided on your Data Sheet, diagram the shape of the paramecium. Use arrows to indicate the direction of movement of the paramecium.

In order to quiet the protozoans (paramecia) for further study, you may prepare traps of cotton fiber or use a drop of methyl cellulose on the slide. Now examine the paramecium under high power. Use your text or charts to identify various parts.

Add a drop of carmine red solution to the slide. These particles will make it easier to observe the cilia. The paramecium will ingest these particles. You should follow their path through the organism. Locate the oral groove and note that cilia also line it.

Locate a food vacuole with carmine red particles. Observe the vacuole for a few minutes to determine movement.

Find out how the paramecium lives in fresh water (a hypotonic solution). Look for clear circles with numerous canals leading to them at either end of the cell. Watch their action for a few minutes.

Finally, add a drop of acetic acid at the edge of the cover glass and observe what happens.

Record your observations by answering Data Sheet questions 9–15.

Make a drawing of the paramecium in the space provided on your Data Sheet. Label all the organelles you can locate. Show the direction of movement of the food vacuoles.

(C) In this section, you will determine whether paramecia respond positively or negatively to chemicals, gravity, heat, and touch.

Response to chemicals. Add a drop of paramecium culture to a slide and focus under low power. Now add a few salt crystals and carefully observe a paramecium.

Wash the slide and add another drop of paramecium culture. Add a drop of vinegar to the culture. Note the response to the vinegar, an acid solution.

Prepare another slide and cover with a cover slip. Place a drop of ink under the edge of the cover glass. Answer Data Sheet questions 16–22.

Response to gravity. Pour paramecium culture into 4 test tubes. Insert stoppers into all the tubes. In order to demonstrate that the response is due to gravity rather than light, cover 2 of the tubes with dark paper. Put a covered and uncovered tube in a vertical position. Put the remaining 2 cultures in a horizontal position. Allow the cultures to remain in their respective positions for 10 minutes. Examine each of the tubes. Answer questions 23–27.

Response to heat. Prepare a slide of paramecia and check under low power to be sure that several animals are present. Add a drop of hot water to one side of the cover glass. Examine the slide under low power at once. Answer question 28.

Response to touch. Put a drop of paramecium culture on a slide for microscopic study. Locate one of the organisms in the field. Put a few strands of spun glass just under the cover slip of a paramecium culture. Answer questions 29–30.

Response to gravity

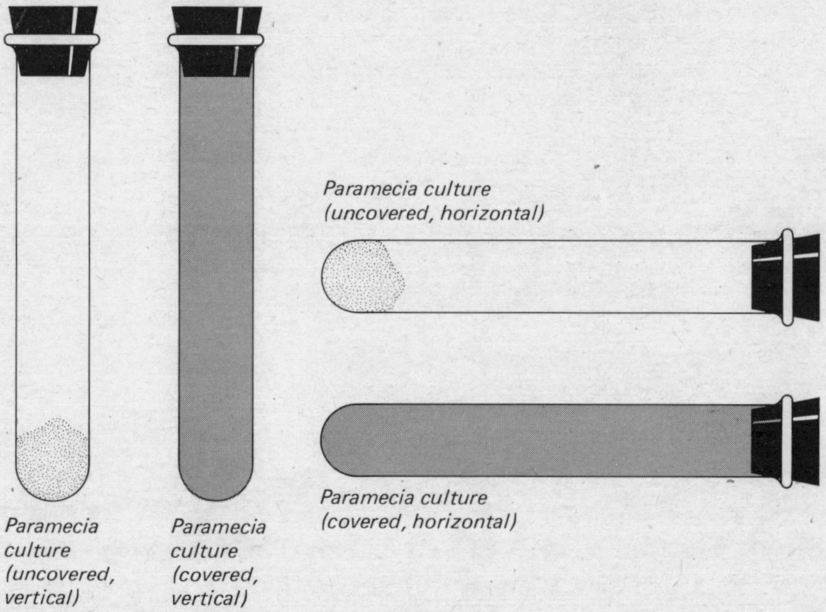

Investigations On Your Own

Collect a number of water samples from different environments. Sea, marsh, lake, and aquarium water can be used. Prepare wet-mount slides of each of these types of water and examine them for microscopic life. What different communities do you find?

Invertebrate Animals

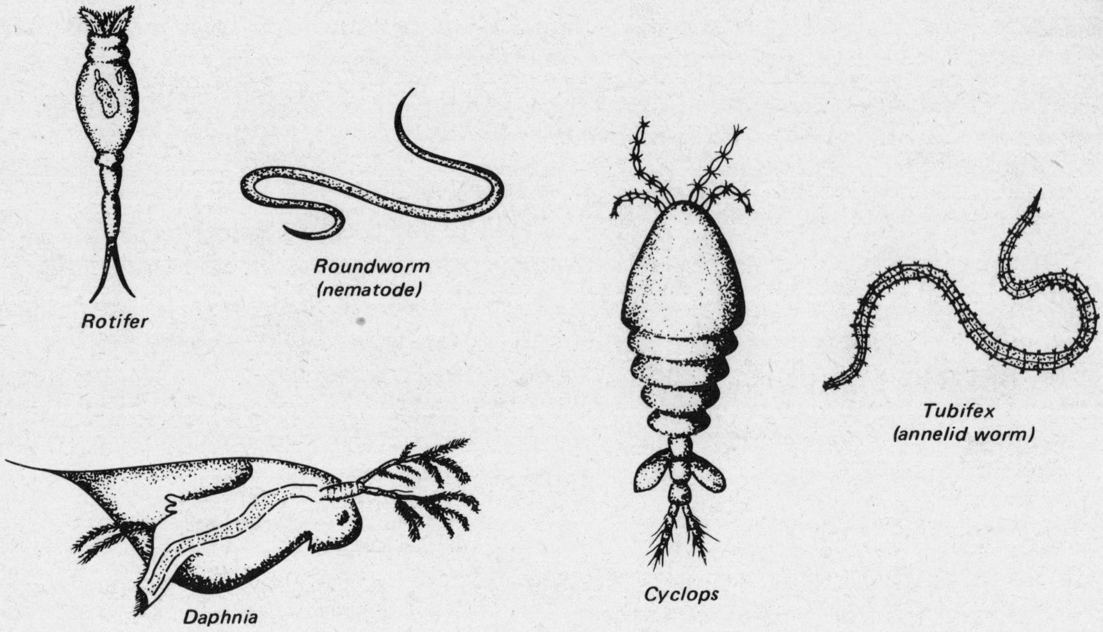

Protozoans

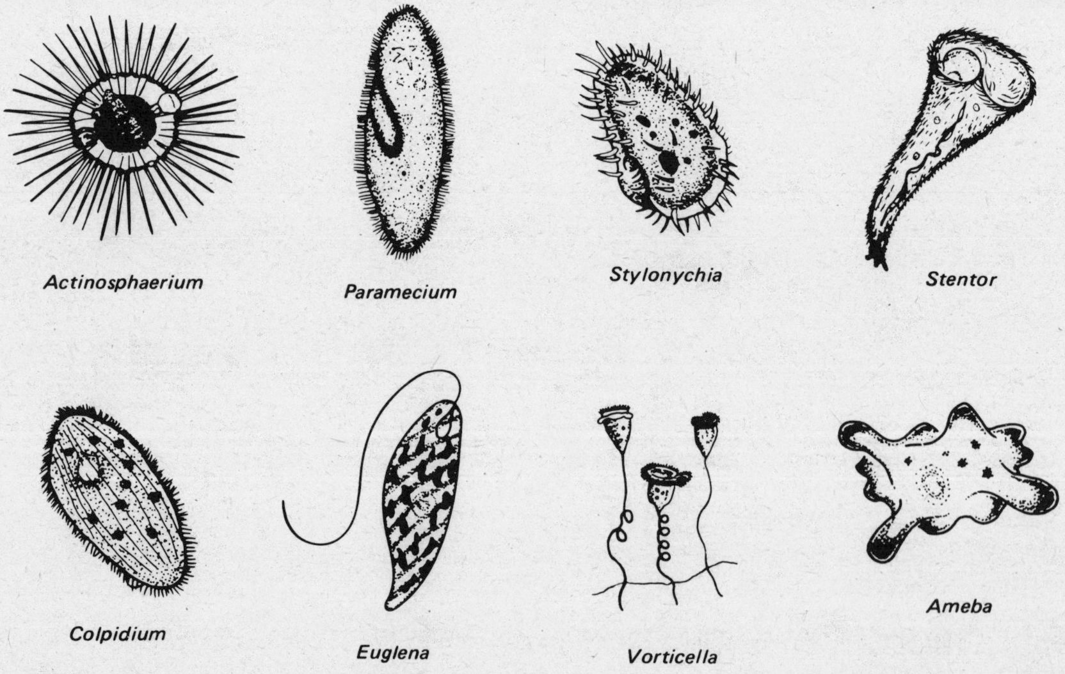

NAME DATE

18 Microscopic Life

DATA SHEET

Analysis

Label these drawings of the ameba and the euglena.

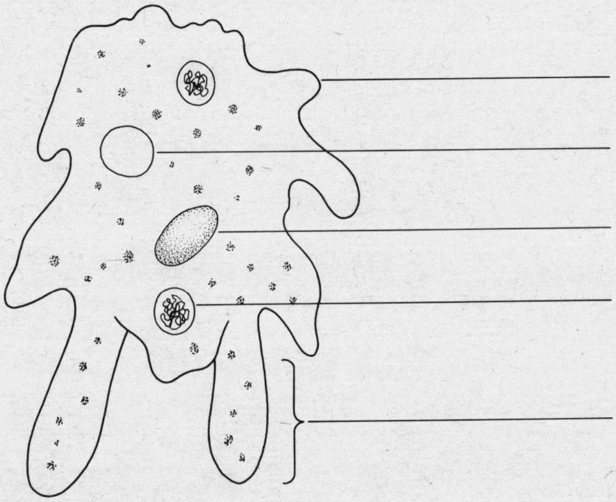

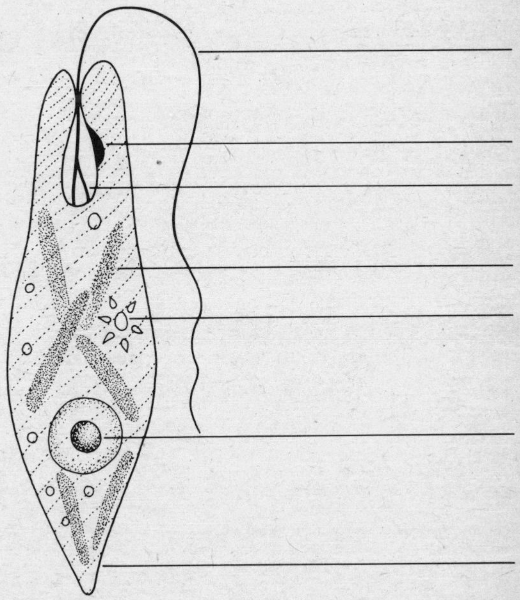

(1) What are some obvious differences between the ameba and the euglena?

(2) How does each organism obtain its food?

Investigation

(A) Refer to the drawings on the opposite page and the illustrations in your text to help you identify the organisms in your pond water sample.

(3) Why is it important to prepare slides of both the muddy water and the clear water from the pond?

In the space below, draw and identify the various organisms you observed.

(4) How can you tell the difference between the protozoans and the plant algae in your sample? ______________________________

____

____

(5) Describe the movement of the protozoans. ______________________________

(6) What happens when the methyl cellulose is added? ______________________________

____

(7) Which organisms are producers? ______________________________

____

(8) How can you determine that they are producers? ______________________________

____

(B) Examine the protozoan — paramecium — as described in the Procedure page.

(9) Observe the movement of the paramecium. Why do you think it moves in this manner? ______________________________

____

(10) Describe the flexibility of the paramecium's shape. Explain your answer. ______________________________

____

____

NAME DATE

(11) Follow the path of the carmine red solution through the paramecium. What would you say is the function of the oral groove cilia? ______

______

(12) In what direction do the food vacuoles move? Explain your answer. ______

______

______

(13) Describe the function of the canals. ______

______

(14) What happens when you add acetic acid to the culture slide? ______

______

(15) What is the probable function of this action? ______

Draw and label all the organelles you located in your paramecium specimen. Use arrows to show the movement of the food vacuoles.

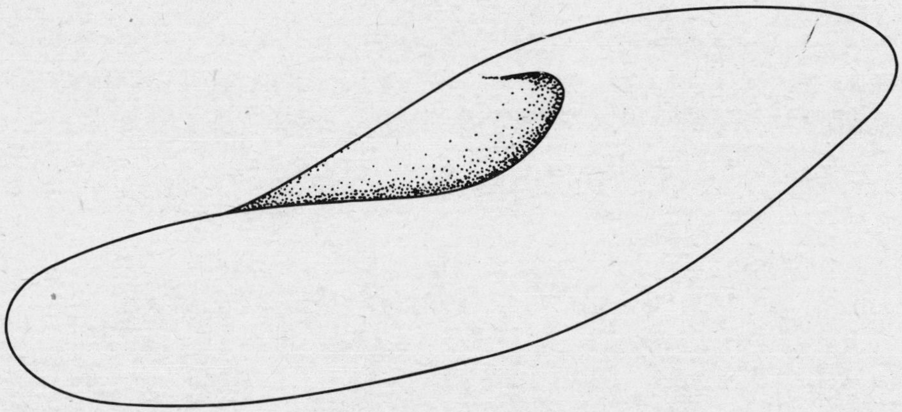

Paramecium caudatum

(C) Determine how paramecia react to chemicals, gravity, heat, and touch.

(16) How does the paramecium react when salt crystals are added to the slide? ______

______

(17) Why does this occur? ______

______

(18) What response does the paramecium make to the vinegar? ______

______

(19) What type of substance is vinegar? ______

(20) Do you think that paramecia tend to live in an acid or alkaline environment?

____

(21) On what evidence do you base your answer to **(20)**? ______________________

____

(22) What happens to the paramecia when ink is placed under the cover slip? ________

____

Answer the following based on the paramecia response to gravity.

(23) Where are the paramecia concentrated in the uncovered tube in the vertical position? ______________________

(24) in the covered tube? ______________________

(25) in the uncovered horizontal tube? ______________________

(26) in the covered horizontal tube? ______________________

(27) Describe the effect of light on the response to gravity. ______________________

____

(28) How do the paramecia react to the heat? ______________________

(29) What is the paramecia's reaction to touch? ______________________

(30) Describe the reaction of the paramecia to a few strands of spun glass. ________

____

Summary

Match the numbered definition with the structure or stimulus on the left.

________	**(a)** pellicle	**1.** positive response
________	**(b)** oral groove	**2.** negative response
________	**(c)** cilia	**3.** ingests food
________	**(d)** food vacuole	**4.** eliminates water
________	**(e)** contractile vacuole	**5.** defense mechanism
________	**(f)** trichocysts	**6.** for locomotion
________	**(g)** heat	**7.** transports food through cytoplasm
________	**(h)** weak acid	**8.** protective surface

(9) What contributions does a consumer make to the pond-water community? ________

____

NAME DATE

Fungi 19

MATERIALS

INVESTIGATION▶

- stock culture of *Penicillium notatum, Aspergillus niger*
- Sabouraud's dextrose agar
- Petri dishes
- inoculating loop
- hand lens
- Bunsen burner
- china marker
- autoclave or pressure cooker
- hot-air oven

OBJECTIVES▶

- To identify and label the structure of common bread mold
- To examine blue-green molds

Problem

What are some forms of mold?

Molds and other forms of fungi grow best under optimum conditions of moisture, darkness, food supply, and temperature. Bread mold and other forms of mold spoil food. Other types of fungi serve as decomposers in our environment. All forms lack chlorophyll and are unable to make their own food. They infect both decaying and living tissues of plants and animals. Enzyme action transforms the materials of the host or food source into a soluble form that is absorbed into tiny threadlike filaments called hyphae.

Bread mold is frequently recognized as a cottony growth on bread and other foods. The drug penicillin is extracted from one species of *Penicillium*. *Penicillium* is found as a green growth upon almost any form of decaying organic matter. *Aspergillus* is a common blue-green mold found on bread, cheese, and decaying fruit.

Procedure

Part I. Analysis

Turn to the Data Sheet. Examine the structure of the bread mold illustrated in the diagram. Refer to your text to help you identify and label the various structures. Fill in the chart and answer questions 1–6.

Part II. Investigation

Using a sterile inoculating loop, transfer a loop of the *Penicillium notatum* culture to the center of the sterile gelled agar as shown in the diagram on the next page. Rotate the loop on the surface of the agar to remove the mold culture. Then, in the same manner, transfer a loop of the *Aspergillus niger* culture to a second Petri dish. Label the bottom of each Petri dish with the name of the mold with which the plate has been inoculated.

Place the inoculated plates in a dark, cool place for a week or more. At the end of this time, use a hand lens to examine the colonies. After removing the lid from the culture dish, examine both cultures under low power of the microscope. Focus on the edge of the colony and locate the fruiting bodies (conidia), which are produced in chains at the tips of certain hyphae called conidiophores. Observe differences and similarities in their structure. Complete the drawings in the space provided on the Data Sheet and fill in the chart.

Investigations On Your Own

(1) Investigate bread mold and blue-green mold that have formed on food products in your home. It is simple to save bread or dairy products until you notice the formation of the mold. Examine these molds and draw conclusions from your observations.
(2) Are bread molds and blue-green molds dangerous to consume? Do some research into this question and design an original investigation that will give scientific proof to your hypothesis. If you are uncertain about your experiment, discuss it with your teacher before carrying out the investigation.

NAME DATE

Fungi 19

DATA SHEET

Analysis

Examine and identify the structures that compose bread mold. On the drawing label: rhizoids (descending hyphae), stolons (transverse hyphae), sporangiophores (ascending hyphae), sporangium, spores, and food source. Use your text to determine the function of each of the hyphae and fill in the chart.

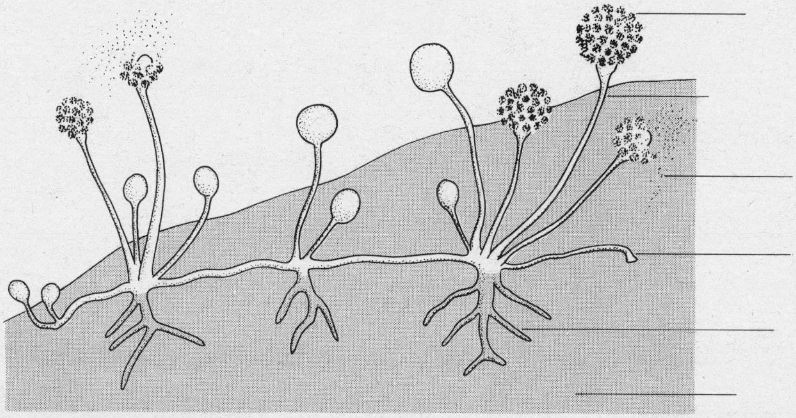

Kind of Hyphae	Function
1.	
2.	
3.	

(1) From what source do bread mold and other molds come? ______________________

(2) How does bread mold begin its activity? ______________________

(3) How do mold plants obtain food? ______________________

(4) How do hyphae that are not in touch with the food source obtain food? ______________________

(5) Why is an odor sometimes detected on molded food? ______________________

____

(6) What advantage is offered by the sporangia developing at the tip of the sporangiophore? ______________________

____

Investigation

In the space provided, draw and label the conidia and conidiophores from each of the molds observed. Complete the chart.

	Penicillium	**Aspergillus**
(1) Colony Form and Color **(2)** Structure of hyphae **(3)** Conidia Arrangement		

Summary

(1) Why do bread molds form on bread that is kept in bread boxes or covered drawers?

____

____

(2) What structural features enable molds to thrive in various environments? ______________________

____

____

(3) How do the reproductive characteristics of molds contribute to their survival?

____

____

____

NAME DATE

Algae 20

MATERIALS

INVESTIGATION▶

- living or preserved cultures of:
  Nostoc
  Oscillatoria
  Gloeocapsa
  Protococcus
  Spirogyra (conjugative and x vegetative forms)
  Odeogonium
  Ulothrix
- microscope
- slides
- cover glasses
- pipette
- dissecting needles
- scissors
- colored pencils
- hand lens

OBJECTIVES▶

- To learn of representatives of various algae phyla
- To observe the structure of different species of blue-green and green algae

Problem

How do various algae differ?

Algae are microscopic plant forms limited largely to a water environment. They are widely distributed in ponds, lakes, streams, rivers, and oceans. The phyla of algae differ from one another on the basis of their cellular structure, photosynthetic pigments, and stored food. The common names given to the phyla of algae refer to the color that is usually most prominent in their cells. Their form is a more reliable basis of classification. Algae may be unicellular or multicellular. Colonies of cells may be formed of loose aggregates of cells or have a definite shape. They may form filaments or plate-like growths. They may be branched or unbranched, attached or free-floating, macroscopic or microscopic.

Procedure

Part I. Analysis

Turn to the Data Sheet and examine the figures of algal forms representative of the major phyla of algae. Use your test as a guide to label the drawings and to determine the name of the alga and its phylum.

Part II. Investigation

As you are ready to examine each form, mount it in a drop of water on a slide. If filamentous, separate the strands with dissecting needles. Add a cover glass and examine under low and high power. Draw each specimen in the space provided on the Data Sheet. Use the words given in **boldface** to label the drawings. Answer questions 1–20.

(a) Obtain a ball of *Nostoc*. Cut a section and mount it on a slide. Mash the section with a dissecting needle to separate the filaments. Notice the many curved and twisting filaments embedded in the *Nostoc* jellylike matrix. Locate swollen, empty cells called **heterocysts**.

(b) *Oscillatoria*, another blue-green alga, is found as a floating or submerged mass of filaments in stagnant water. Different species of *Oscillatoria* vary greatly in size and range in color from bright blue-green to nearly black. Prepare a slide. If the specimen is living, watch for a slight swaying motion near the ends of the filaments. The cell at the end of the filament is called the **apical cell**. Various species of *Oscillatoria* differ in the shape of this cell. Determine whether you have more than one species. Locate empty, collapsed cells known as **concave cells**. Draw several filaments of *Oscillatoria* in the space provided.

(c) *Gloeocapsa* is a primitive blue-green alga, often found on moist flower pots and wet rocks. A **cell** secretes a slimy **sheath** that encases it. As cells divide, each new cell secretes a sheath within the old one. Examine a drop of *Gloeocapsa* on a slide. Reduce the light entering the objective and study the colony.

(d) *Protococcus*, one of the most common of the unicellular forms of green algae, can be found as a green film on damp walls, rocks, fence posts, and the bark of trees. Examine the growth of *Protococcus* on bark with a hand lens. Scrape some algae into a drop of water on a slide. Draw a single **cell** and several colonies. Shade the **chloroplast** green.

(e) *Spirogyra* is a common, unbranched, filamentous alga that forms a green growth on the surface of ponds and streams. Examine the cells and locate bodies of stored food on the **chloroplasts**. These are the **pyrenoids**. The **cytoplasm** lies in a dense layer around the wall and extends through the cell in strands. The **nucleus** is embedded in the strands of the cytoplasm. The large cavities are **vacuoles**. Observe the **gelatinous sheath** surrounding the filaments. Make a drawing of three cells in a filament showing how the cells are joined. Shade the chloroplast green and the pyrenoids red.

(f) Conjugating *Spirogyra*. During conjugation, the entire content of one cell passes through the **conjugation tube** and unites with the content of the opposite cell. The contents, in this situation, behave as **gametes**. The fusion body resulting from the union of the two gametes is a **zygote**. Locate two filaments in the process of conjugation.

(g) *Ulothrix* is a filamentous green alga whose basal cell may be modified to a **holdfast cell**. The alga may be attached to rocks or be free-floating. Locate the holdfast cell, **chloroplasts**, **nucleus**, and **pyrenoids**.

(h) *Oedogonium* is a filamentous green alga usually attached by a **holdfast cell** when young but free-floating when mature. Observe similarities and differences in the cells. The large, rounded **oogonial cells** produce the egg cell. The small, short **antheridial cells** produce the sperm cells.

Investigations On Your Own

(1) You may be interested in studying diatoms found in fresh and salt water. Diatoms occur singly or in colonies and have a wide range of shapes and colors. Prepare a wet mount of a diatom culture and examine the structure, forms, and colonies of these organisms. What colors do you observe?

(2) A study of marine algae could be of particular interest to those living in coastal areas. These algae are usually brown and red. The brown algae have multicellular sex organs and a complex vegetative body. The red algae lack flagellated cells and contain a red pigment. The thallus is often multicellular and may be simple or branched. Your culture of marine algae may be fresh or preserved. It should include *Fucus* and *Chondrus crispus*. Examine these specimens and others you are able to obtain. Locate the **holdfast**, **stipe**, and **frond**.

(3) Collect pond water from various sources. Try to obtain samples from different depths and conditions. Observe under low and high power. Attempt to identify the algae you have studied in this Investigation. A key or various guides to algae may be used to identify the forms you have collected.

NAME ______________________________ DATE ______________

Algae 20

DATA SHEET

Analysis

Identify each of the alga forms below. Use your text as a guide to label the drawings and to determine the name of the alga and its phylum.

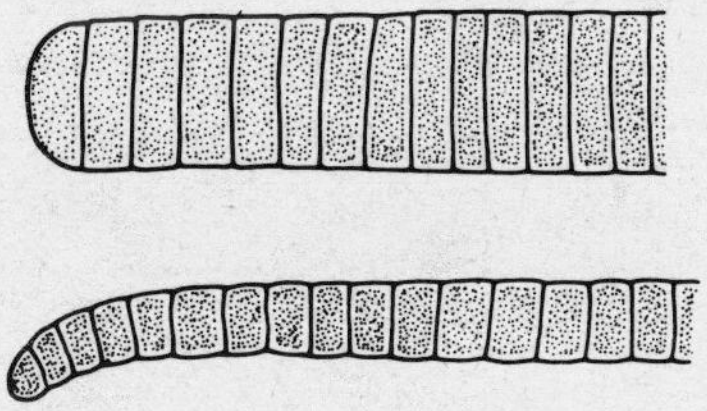
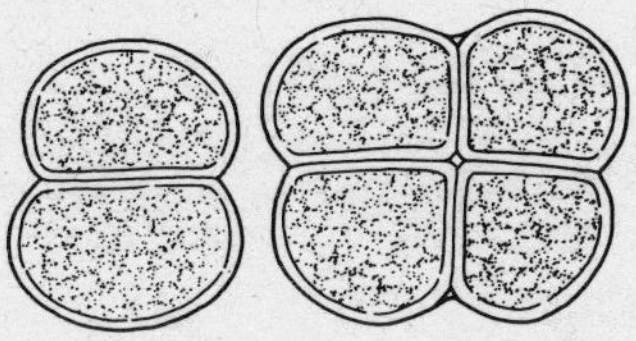
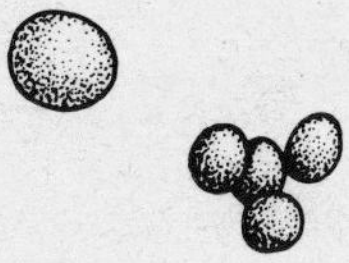

Name ______________ ______________ ______________

Phylum ______________ ______________ ______________

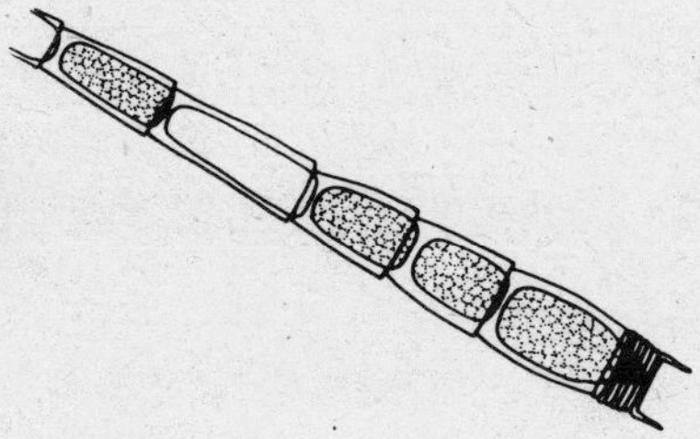

Name ______________ ______________ ______________

Phylum ______________ ______________ ______________

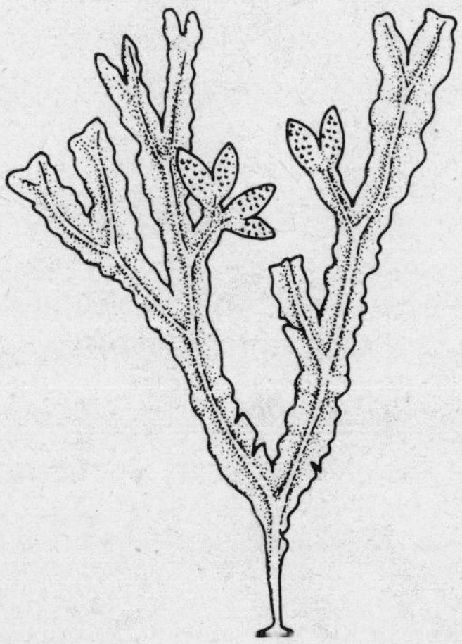

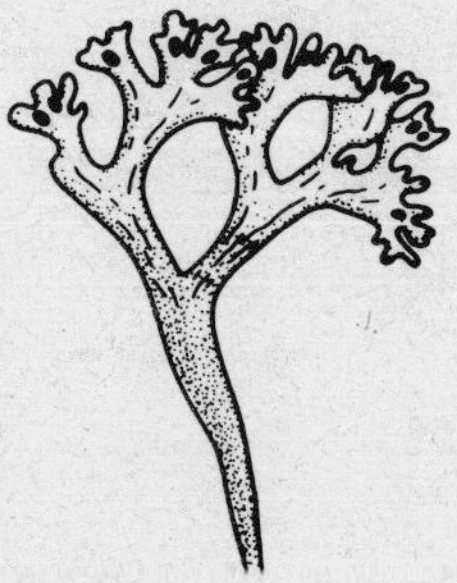

Name ______________ ______________ ______________

Phylum ______________ ______________ ______________

Investigation

After examining the forms of algae, draw each in the space provided and label the drawings using the structures given in **boldface** in the directions. Answer the questions.

Name ______________________

Phylum ______________________

Name ______________________

Phylum ______________________

Name ______________________

Phylum ______________________

(1) Describe the distribution of heterocysts in the filaments of *Nostoc*. ______________________

____

____

(2) What is the function of the heterocysts? ______________________

____

(3) How do concave cells function in *Oscillatoria*? ______________________

____

(4) List the ways in which *Gloeocapsa* is similar to bacteria. ______________________

____

(5) How is it different? ______________________

Name ______________________

Phylum ______________________

Name ______________________

Phylum ______________________

Name ______________________

Phylum ______________________

Name ____________________

Phylum ____________________

Name ____________________

Phylum ____________________

(6) With regard to its growth on the bark of trees, why is *Protococcus* not considered a parasite? ____________________

(7) Describe the walls between the cells of *Protococcus* that are joined. ____________________

(8) Are the cells in a filament of *Spirogyra* cell the same size and shape? ____________________

(9) How many cells constitute a *Spirogyra* organism? ____________________

(10) Do all of the chloroplasts spiral in the same direction in the cells of a filament? ____________________

(11) Is the number of pyrenoids uniform for each cell of a filament? ____________________

(12) How would the sheath aid in identification of *Spirogyra* if you were collecting it from a pond or stream? ____________________

NAME DATE

(13) How can you tell when *Spirogyra* cells are conjugating? ______________________

____

____

____

(14) Is the movement of the cell content all in one direction? ______________________

(15) Do all the cells in the two filaments conjugate? ______________________

(16) Describe the zygote. ____

____

(17) How does the holdfast cell of *Ulothrix* differ from other cells of the filament?

____

____

(18) Describe the chloroplast. ____

____

(19) In *Oedogonium*, how do the cells of the filament differ from each other? __________

____

____

(20) How does reproduction in *Oedogonium* differ from that in *Spirogyra*? __________

____

____

____

Summary

Next to each term at the left, write the numbers of the algal forms that apply:

(1) blue-green algae ______________________

(2) chlorophyta ______________________

(3) heterocysts ______________________

(4) gelatinous matrix ______________________

(5) filamentous ______________________

(6) swaying motion ______________________

(7) concentric sheaths ______________________

(8) damp tree bark ______________________

(9) spiral chloroplasts ______________________

(10) pyrenoids ______________________

(11) conjugation ______________________

(12) holdfast cell ______________________

(13) concave cells ______________________

1. *Nostoc*
2. *Gloeocapsa*
3. *Oscillatoria*
4. *Protococcus*
5. *Ulothrix*
6. *Spirogyra*
7. *Oedogonium*

NAME DATE

Mosses and Ferns 21

MATERIALS

INVESTIGATION▶

- fresh or preserved moss
- plants bearing sporophytes
- fresh fern frond bearing sori
- prepared slide of fern gametophyte
- 3 × 5 index card
- straight pin
- dissecting microscope
- textbook or charts
- glycerin
- dropper

OBJECTIVES▶

- To study the life cycle of mosses and ferns
- To learn the meaning of alternation of generations
- To examine structural features of mosses and ferns

Problem

How can plants reproduce without seeds?

The familiar clumps or "carpets" of green moss plants are the leafy shoots or gametophyte generation. Although the leafy shoots may appear to have a stem and roots, they are not considered as such because they lack conducting and supporting tissues found in these organs of higher plants. Moss plants, therefore, grow close together for support. At certain times of the year, sporophytes may be seen growing from certain moss plants.

Ferns occupy many of the same habitats as mosses. The familiar fern plant is the sporophyte generation. Unlike the moss sporophyte, it is free-living and independent. Both mosses and ferns exhibit alternation of generations, in which a spore-producing plant, the sporophyte, alternates with a gamete-producing plant, the gametophyte. Ferns are able to complete their life cycle without the production of seeds.

Procedure

Part I. Analysis

(A) Study the diagram of the moss life cycle on the Data Sheet. You may want to use your textbook as a guide. Identify the *gametophyte,* which is the familiar moss plant. The rootlike structures at the base of the leafy shoot are *rhizoids,* which function to absorb water and minerals.

At the tip of the leafy shoot are located either male *antheridia,* which produce motile sperm cells, or female *archegonia,* which produce the egg cells. When a clump of moss plants becomes mature with gametes and is covered with a film of water, the sperm swim to the egg. Fertilization is accomplished in the archegonium. The resulting *zygote* becomes the first cell of the new, dependent, *sporophyte* generation. Mitotic divisions of the zygote within the archegonium form the *embryonic sporophyte.* Further division results in the growth of the visible sporophyte.

The sporophyte generation consists of a *seta* (stalk) and a *capsule*. At the top of the capsule is a lidlike covering, the *operculum*. As *spores* mature within the capsule, the operculum will fall away and the spores will be released. Under favorable conditions a spore germinates into a threadlike structure, the *protonema*. The protonema forms a young gametophyte plant that eventually develops into the mature gametophyte.

On the Data Sheet, label the life cycle of the moss plant. Answer questions 1–5.

(B) Study the diagram of the fern life cycle on the Data Sheet. The familiar fern plant is the *sporophyte* generation. The *fronds* (leaves) of the fern grow from a horizontal, underground stem, the *rhizome*.

The fronds first emerge from the ground as *fiddleheads*. Each consists of a stalk and blade divided into leaflets, or *pinnae*. When certain of the fronds mature, small dots known as *sori* (sing. sorus) are produced on the lower surface. On the single pinna shown in the life cycle, identify a sorus containing *sporangia*. When spores are released and land in a favorable location, they germinate and develop into *gametophytes*. The gametophyte is called a *prothallus*. The prothallus measures about 1 cm in diameter. The prothallus develops *antheridia* and *archegonia*, which produce sperm and eggs. When moisture is present, the sperm swim to the egg and fertilization is accomplished. Soon a *young sporophyte* develops from the gametophyte.

Turn to the Data Sheet and label the fern life cycle. Answer questions 6–8.

Part II. Investigation

(A) Examine a moss plant bearing a sporophyte plant. Locate the parts labeled in the Analysis section. Remove a leaf, make a wet mount, and examine all areas of the leaf under low power. Note the arrangement of the cells and any cell organelles that may be present. On the Data Sheet answer questions 9–12.

On the moss plant, observe that the seta attaches the sporophyte to the gametophyte. Spores are released from the capsule through the *peristome teeth*. To observe the peristome teeth, prepare a 3 × 5 card in the manner shown in the figure. Use a straight pin to poke a hole through the card. Pull the seta through the hole until the capsule is held snugly in an upright position. Remove the operculum if it is present. In the space provided on the Data Sheet, draw a side and top view of the capsule. Label the structures given on the Data Sheet. Answer questions 13–15.

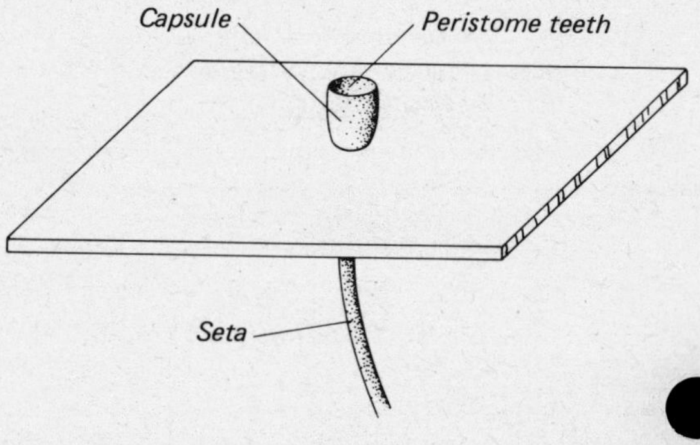

(B) Examine a prepared slide of a fern gametophyte or prothallus under low and high power. Locate the egg-producing *archegonia* at the base of the *apical notch*. Among the rootlike *rhizoids* are the sperm-producing *antheridia*. Observe the structure of the rhizoids. Turn to the Data Sheet and on the diagram of the prothallus label: antheridia, archegonia, rhizoids, apical notch.

Examine a fresh fern frond that has formed sori. Cut around a sorus, prepare a wet mount, and examine it under low power. Remove the coverlike *indusium*, if present, and observe the helmet-shaped *sporangia*. The thickwalled cells going around the sporangium are those of the *annulus*. The annulus functions to disperse *spores* under favorable environmental conditions.

Examine the sorus under low power. Observe the cluster of sporangi and the cells composing the sporangium. On the drawing of the sorus label: *sorus, indusium, cells of pinna, sporangium, annulus, lip cells, spores*. After labeling the drawing, add a drop of glycerin to the side of the sorus and watch for changes in the sporangium. If no change occurs, add glycerin until a change occurs. Answer questions 16 and 17.

Investigations On Your Own

(1) Make a collection of moss gametophytes bearing stalks and capsules. Look in locations where there is moisture and reduced light. Record the location, date, and habitat in a notebook. Use identification keys to trace your specimens to their particular genus. Leaf and capsule structures are often used for identification purposes.
(2) Make a collection of fern fronds. Fronds can be pressed and saved for later study in the laboratory. Identify the fronds by means of a key. General form and structure of the sporangia and indusia and distribution of the sori are characteristics used in their identification.

NAME DATE

Mosses and Ferns 21

DATA SHEET

Analysis

(A) Examine the stages in the life cycle of the moss plant. Label the following structures: *rhizoids, leafy shoot, archegonium, antheridium, egg, sperm, zygote, embryo of sporophyte, seta, capsule, operculum, spores, protonema, young gametophyte*. Indicate the *gametophyte generation* and the *sporophyte generation*.

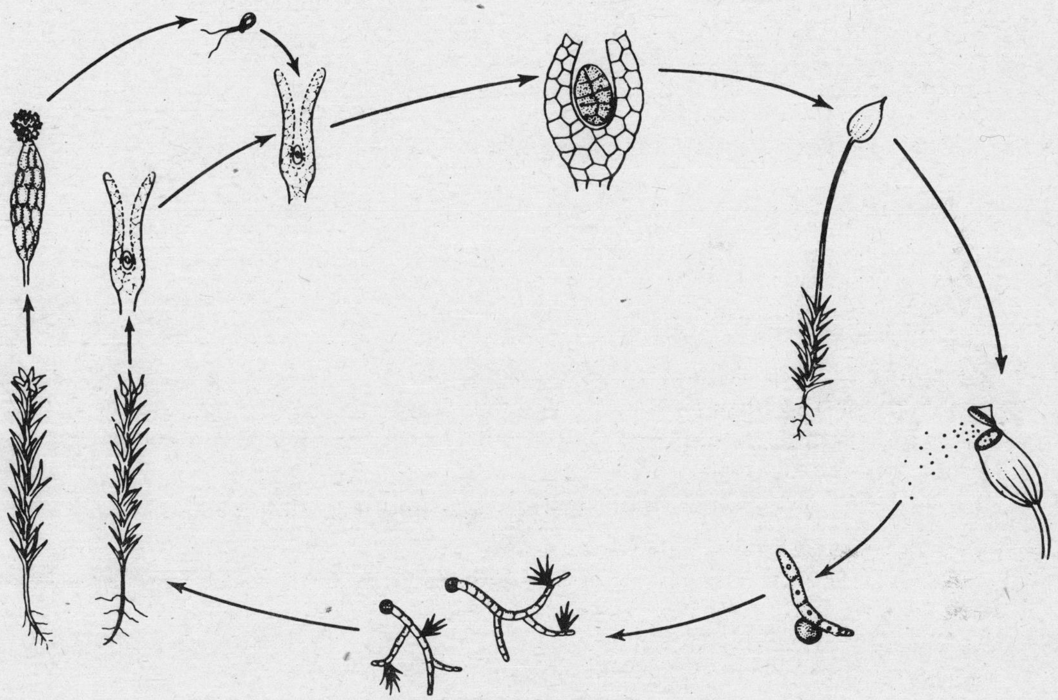

Life cycle of a moss plant

(1) Explain why mosses are not found growing in locations having little or no moisture.

(2) Why is the seta and capsule known as the sporophyte generation?

(3) From what cell did the sporophyte develop?

(4) Explain why the sporophyte might be present only on certain gametophyte plants.

(5) What may be observed about the sporophyte in the diagram that indicates that it is nutritionally dependent on the gametophyte? ______

______

______

(B) On the diagram of the fern life cycle, label: *rhizome, frond, pinna, sorus, sporangia, spores, prothallus, archegonium, antheridium, egg, sperm*. Also indicate *sporophyte generation* and *gametophyte generation*.

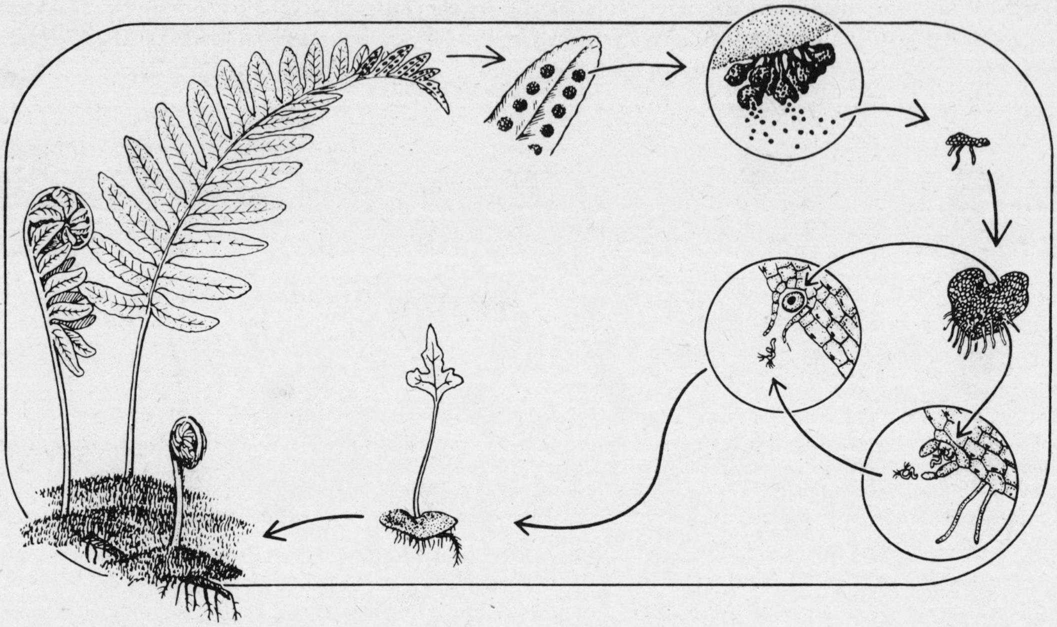

(6) Compare the sporophytes of the moss and fern. ______

______

______

(7) With what structure of a moss does a fern sorus compare? ______

______

(8) Describe the number and position of the sori. ______

______

Investigation

(A) Examine the stem axis and the "leaves" it bears. Remove a leaf and prepare a wet mount. Examine it under low power and high power.

(9) Describe the arrangement and attachment of the leaves on the stem axis. ______

______

(10) Suggest an adaptive value of the leaf arrangement. ______

______

NAME DATE

(11) Of what adaptive value is a greater thickness along the midrib? ______________________

____

(12) How are the chloroplasts arranged in a single cell? ______________________

____

Follow the directions as outlined in the Procedure. In the space provided, draw a side view and a top view of a capsule. On the side view label: *seta, capsule, operculum*. On the top view label: *wall of capsule, peristome teeth*.

(13) Describe the top of the capsule. ______________________

____

____

(14) What is the function of the peristome teeth? ______________________

____

____

(15) How is spore dispersal accomplished? ______________________

____

(B) Examine the prepared slide of a fern prothallus under low power and locate the male and female sex organs. On the figure of the prothallus label: *rhizoids, antheridium, archegonium, apical notch*.

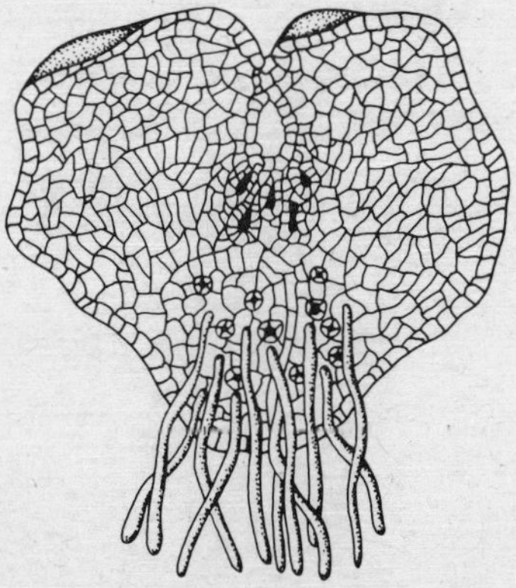

Prothallus

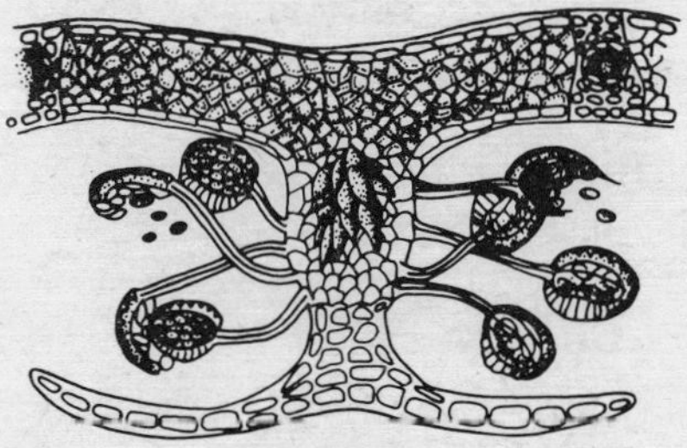

sorus

(16) Where did the cell separate as the sporangium opened? ______________________

____

(17) Describe the change in the cells of the annulus after glycerin was added. ______

____

____

____

____

Summary

(1) Explain how moss and fern plants exhibit alternation of generations. __________

____

____

____

(2) Summarize your observations of mosses and ferns by filling in the blanks of the following chart.

	Mosses	Ferns
(1) Produces eggs		
(2) Produces sperm		
(3) Sporophyte independent (yes/no)		
(4) Structure producing spores		
(5) Spores germinate into:		
(6) Inconspicuous generation		
(7) Conspicuous generation		
(8) Medium for fertilization		
(9) Agent for spore dispersal		
(10) First cell of sporophyte		
(11) First cell of gametophyte		
(12) General habitat		

NAME DATE

Seed Plants 22

MATERIALS

INVESTIGATION▶

- Fresh or preserved specimens of:
  pollen (staminate) cones
  1st year seed (ovulate) cones
  2nd year seed cones
  3rd year seed cones
- dissecting microscope
- microscope
- scalpel or razor blade
- forceps
- dissecting needle

OBJECTIVES▶

- to identify the organs of a seed plant
- to examine the cones involved in the life cycle of a typical gymnosperm

Problem

What are seed plants?

Seed plants have become the most abundant form of vegetation on earth due to the development of highly specialized roots, stems, and leaves. In addition, the flowers of angiosperms and the cones of gymnosperms provide the most highly specialized forms of sexual reproduction in the plant kingdom. Gymnosperms are very successful land plants, as seen by the large areas occupied by members of this group. Only the flowering plants are more successful and occupy a larger portion of the land surface. You will study angiosperm structure in greater detail in later Investigations.

In the gymnosperms and angiosperms, both the gametophyte and sporophyte are modified to accomplish seed formation. In these groups, it is the seed containing the embryo plant that becomes the dominant stage in the life cycle.

Procedure

Part I. Analysis

Turn to the Data Sheet. Study the diagram of the seed plant. With the aid of your textbook, determine the functions of each of the organs that compose the seed plant. Label the drawing and answer questions 1–14.

Part II. Investigation

(A) Examine a branch of a gymnosperm-bearing pollen cones. Determine the arrangement of the cones. Remove a single cone and examine it under a dissecting microscope. The parts of an individual cone are similar to the sporangia seen in fern plants. Locate the *central axis* to which many *spore leaves* are attached. These structures bear two saclike structures that form the *pollen grains* and are known as *pollen sacs*.

Pollen grains are produced from microspores through the process of meiosis. Dust some pollen onto a slide and examine it under a microscope. Notice the "wings" extending from the pollen grain. Close examination should reveal two nuclei within the mature pollen grain. One, the tube nucleus, controls the development of the pollen tube, which functions in delivery of the second, the sperm nucleus, to one of several eggs formed in the female cone. Turn to the Data Sheet and draw a pollen cone and several pollen grains in the space provided. Label the structures. Answer questions 15–17.

(B) The development of seeds in a pine tree spans a period of three years following pollination. Cones, therefore, may be examined in each year of development. Examine a branch bearing the young, first-year, seed or ovulate cone. Compare its location on the branch, its size and appearance, to the pollen cones. Using a scalpel, make a longitudinal slice through the middle of the cone, as shown in the diagram. Observe that it is composed of a *central axis* to which are attached *scales* that increase in size as the cone matures. The ovules or seeds develop on the upper side of the scales after fertilization is accomplished. On the Data Sheet draw a first-year cone in the space provided. Label the structures given in boldface. Answer questions 18–19.

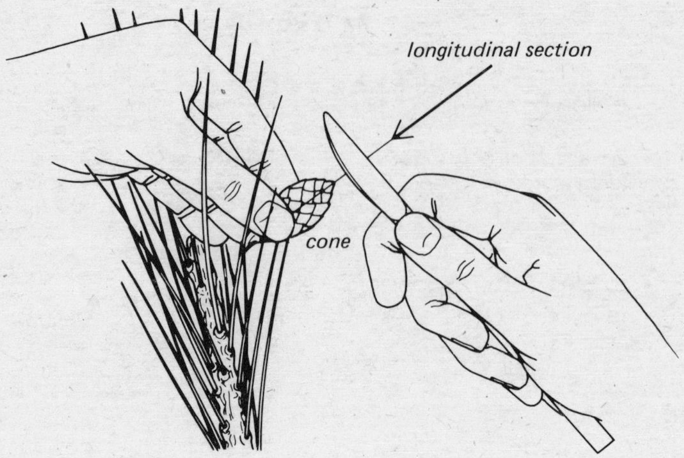

(C) Examine a branch bearing a cone in its second year of development. Determine the differences between this second-year cone and a first-year cone. Carefully remove a single scale. Locate and determine the number of ovules or *seeds* at the base of the *scale*. Make a longitudinal slice in the cone and look for differences in structure between the first-year cone and this cone. On the data sheet, draw a longitudinal section and label the structures. Answer questions 20–21.

(D) After a year's delay in fertilization, a long period is necessary for the maturing of the ovules or seeds. Examine a mature, woody, third-year cone. Examine closely the upper surface of the *cone scales* and notice how the *seeds*, with their *winglike attachment*, are borne on the cone scale. Speculate on how the seeds are probably dispersed. On the Data Sheet, draw a third-year cone and a seed. Use the structures given as labels. Answer questions 22 and 23.

Investigations On Your Own

(1) Design an investigation to determine how fast the vegetative organs of a seed plant grow. Begin with bean seedlings and use India ink to mark the vegetative organs so as to indicate where and how fast growth has taken place. Keep accurate records of the length of the stem, the number of leaves produced, total leaf surface area, and, at the end of the investigation, the total length of the roots. Summarize your findings with a graph. Discuss the project with your teacher for additional directions.

(2) Collect short branches of different gymnosperm species growing in your area either as part of a planted landscape or as a naturally growing species. Note similarities and differences in the habit of growth, how leaves are borne (singly, in fasicles, flattened scales, or awl-shaped) on the branches. Make a cross-sectional sketch of the needles and a sketch of the leaves or fasicles. If cones are present, make a drawing of the third-year cone. Summarize your observations in a chart and with a series of drawings.

NAME DATE

22 Seed Plants

DATA SHEET

Analysis

On the figure label:
roots,
stem,
leaves,
flowers.

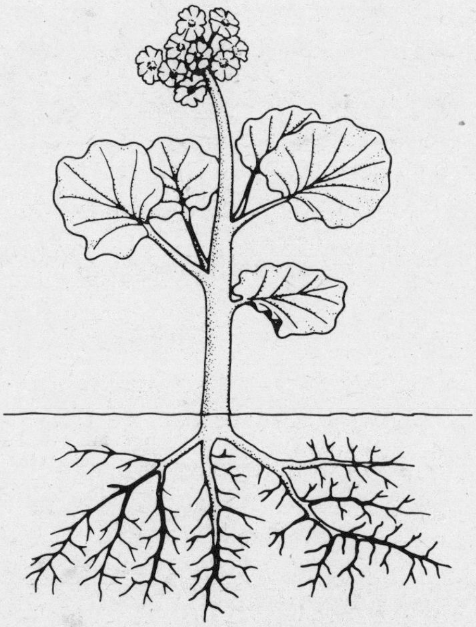

(1) What organs of the seed plant may be observed? ______

(2) What organs are considered to be vegetative? ______

(3) What does it mean for an organ to be vegetative? ______

(4) What characteristic may be observed that indicates the anchoring function of the roots in the soil? ______

(5) What obvious plant organ is attached to the stem? ______

(6) Stems of many plants are green. What pigment is present?

(7) Name another function that may be carried out in the stem.

(8) How do water and minerals get to the leaves from the roots?

(9) How do materials move from the leaves to the roots?

(10) What is the usual color of the leaves?

(11) What is the principal function of the leaves?

(12) What other functions are carried out by a leaf?

(13) Why is a flower not considered a vegetative organ?

(14) How do flowers function to reproduce the plant?

Investigation

In the spaces provided, draw a pollen cone and a pollen grain. On the pollen cone label: *pollen cone, central axis, spore leaves*, and *pollen sacs*. On the pollen grain label: *pollen grain, wings, tube nucleus*, and *sperm nucleus*.

(15) How are the pollen cones arranged?

(16) Suggest a function of the wings on the pollen grain.

NAME DATE

(17) Explain the chromosome number you would expect a pollen grain to have. ______

Make a drawing of a *first-year cone* in longitudinal section and label: *central axis, ovulate scale*.

(18) Suggest a reason why the ovulate cone is located at the tip of the branch. ______

(19) Compare similarities and differences in the pollen and ovulate cone. ______

Draw a longitudinal section of a *second-year cone* in the space provided and label: *cone scale, seeds*.

(20) What differences can be noted between the second-year cone and a first-year cone?

(21) How many seeds can be located at the base of each scale? ____________

____

Draw a *third-year cone* in the space provided and label: *cone scale, seeds*.

(22) Describe the changes that have occurred in the cone since the second year. ______

____

____

(23) How are seeds probably dispersed from the third-year cone? ____________

____

Summary

In the space provided, indicate which organ(s) of the seed plant perform the function indicated.

(1) ____________ conduct water and minerals to upper plant parts.

(2) ____________ are the principle organs of photosynthesis.

(3) ____________ serve in the reproductive process of the plant.

(4) ____________ serve to anchor the plant.

(5) ____________ produce leaves.

(6) ____________ store food substances.

(7) ____________ exchange gases between the plant and the atmosphere.

(8) ____________ absorb water and minerals.

(9) ____________ produce the seeds of the next generation.

(10) ____________ conduct water and minerals up and down the plant.

(11) ____________ display leaves to light.

(12) ____________ function in the process of transpiration.

NAME DATE

Leaves 23

MATERIALS

INVESTIGATION▶

- fresh or thawed leaves of spinach
- droppers
- test tube rack
- filter paper
- ethyl alcohol
- solvent (1 part diethyl ether to 3 parts petroleum ether — CAUTION: FLAMMABLE)
- hot water bath
- 300-mL beaker
- forceps
- cork
- 15-cm test tube
- pin or tack
- paper toweling
- glass stirring rod

OBJECTIVES▶

- To determine how internal leaf structure makes a leaf the primary photosynthetic organ
- To determine, by extraction, which pigments are present in a leaf

Problem

How does leaf structure relate to its function?

The green leaf is the chief site of the photosynthetic process. You have already studied the role of carbon dioxide and light energy in photosynthesis. In this Investigation you will study the structures of the leaf that are involved in the process. The green leaf is broad, thin, flat, and arranged on a stem so that its surface is exposed to light.

All of the tissues of the leaf are specialized for a particular role in photosynthesis. As you study the leaf tissue in cross section, you will become familiar with the specialization of leaf structures for absorption of light, conduction of materials, absorption of carbon dioxide, and production of sugar.

Procedure

Part I. Analysis

Turn to the Data Sheet and study the figure of a leaf cross section. Observe that the leaf is composed of three tissues: epidermis, mesophyll, and conducting tissue. The *upper epidermis* is covered by a waxy layer called the *cuticle*. The mesophyll is the largest area of the leaf and is composed of two layers. The first layer is made of *palisade cells,* which lie just below the upper epidermis. Locate some *chloroplasts* in the palisade cells.

Situated below the palisade cells are the loosely arranged cells of the *spongy layer*. Locate chloroplasts in the spongy cells. Observe the numerous *air spaces* located between the spongy layer cells. In an actual leaf, the spongy layer is penetrated by numerous veins. The largest of these can be seen externally with the unaided eye.

Observe that the vein shown is composed of two types of cells. The upper part is the *xylem,* composed of empty, thick-walled cells. The thin-walled cells that form a cluster below the xylem cells are the *phloem cells.* Study the lower epidermis and observe the tiny pores with small, rounded cells on either side. The pores are the *stomata* and the surrounding cells are the *guard cells.* Label the figure on the Data Sheet with the labels given in boldface. Answer questions 1–16.

Part II. Investigation

Place a large spinach leaf in a beaker of boiling water. Allow it to boil for about one minute. Remove the leaf with forceps and pat dry with a paper towel. Tear the leaf into pieces. Insert the pieces in a 15-cm test tube about one-third full of ethyl alcohol. Place the test tube with the leaf pieces in a hot-water bath until the alcohol begins to turn green. Remove the test tube from the water bath. Crush the leaves with a glass stirring rod. Continue heating and crushing until the solution becomes a dark-green color.

Cut a strip of filter paper to just fit in a 15-cm test tube. Cut a point at the end of the strip. Draw a faint pencil line 2 cm from the point. Attach the filter paper to a cork stopper as shown in the figure below. Use a glass rod to trace a thin line of the extract over the pencil line 2 cm from the pointed end of the filter paper. Allow the extract to dry. To build up a quantity of pigments, repeat this procedure about 20 times. Before proceeding, be certain all flames are extinguished. Place 5 mL of solvent in a test tube.

Attach the filter paper strip with a pin or tack to a cork stopper. Adjust the paper strip so that the point of the filter paper just comes in contact with the solvent. Allow the solvent to ascend the chromatogram for 15 minutes. When the solvent has reached the top of the strip, remove it and allow it to dry. Notice the faint bands of color. From the top may be seen: carotenes (orange), xanthophylls (yellow), chlorophyll b (yellow-green), chlorophyll a (blue-green), and anthocyanin (red).

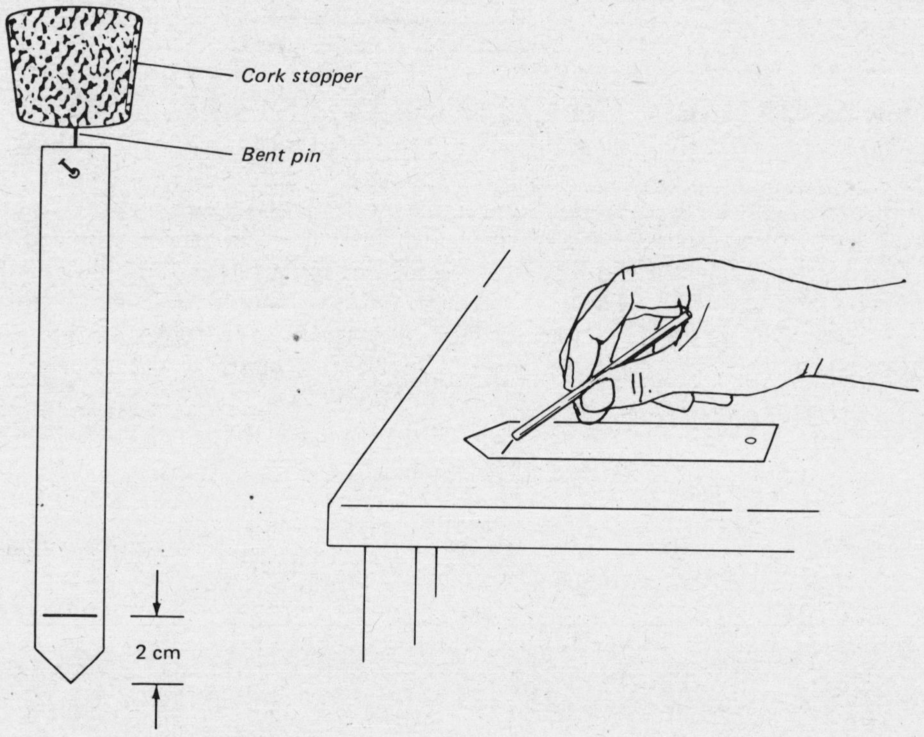

A B

Fill in the diagram on the Data Sheet and answer questions 17–21.

Investigations On Your Own

(1) Select fresh leaves (tulip, geranium, or Kalanchoe are good). Remove a small area of the lower epidermis by tearing through the blade, twisting slightly as you tear. The epidermis will appear as a thin, transparent skin. Using a razor blade, cut off a small portion of this skin. Mount it in a drop of water and add a cover glass. Examine it under low power. Compare the epidermal cells and stomata of several different kinds of leaves. Make drawings of each. Label the drawings where appropriate.

(2) Obtain leaves from other plants, particularly those having an obvious red pigmentation. Repeat the procedure of Part II and determine what pigments are present. Summarize your findings in a series of labeled drawings.

NAME ____________________ DATE __________

23 Leaves

DATA SHEET

Analysis

On the figure of the leaf cross section, locate the cells and tissues that compose a leaf. Label: *cuticle, upper epidermis, palisade cells, chloroplasts, spongy cells, vein, xylem cells, phloem cells, lower epidermis, air space, stoma, guard cells*.

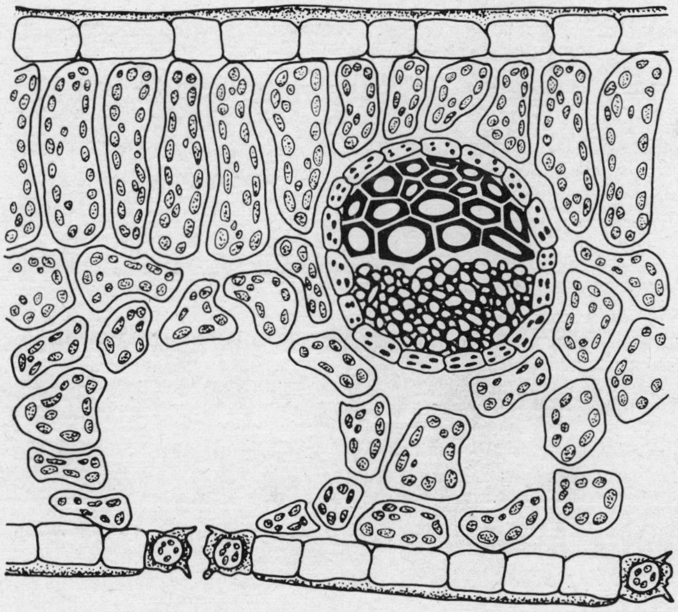

Cross section of a leaf

(1) How many cells thick is the upper epidemis? ____________________

(2) Are any chloroplasts present? ____________________

(3) Suggest a function of the cuticle. ____________________

____

(4) Describe the cells of the palisade layer and their orientation to the upper epidermis.

____

____

____

(5) Suggest a function of the palisade cells. ____________________

____

(6) Why is the shape of the palisade cells important? ______

______

______

(7) Of the two mesophyll layers, which is most compact? ______

(8) Are chloroplasts as numerous in the spongy cells as they are in the palisade cells?

______

(9) Account for any difference you find. ______

______

______

(10) On the basis of their relationship to other leaf tissues, what is the function of the air spaces? ______

______

______

(11) How can you tell a vein from other structures in the leaf? ______

______

______

(12) Suggest two functions of the xylem cells. ______

______

(13) What is the function of the phloem cells? ______

______

(14) How many cell layers compose the lower epidermis? ______

(15) What is the relationship between the stomata and the air spaces of the spongy tissue? ______

(16) Suggest a function of the stomata. ______

______

______

______

NAME DATE

Investigation

On the diagram, accurately sketch the bands of color as obtained on your chromatogram. Use appropriate colors to indicate the color of pigments and identify each band.

C

(17) Did the leaf pigments dissolve better in the water or in the alcohol? ____________________

(18) Why did the separation of pigments occur as it did? ____________________

____

____

(19) What pigments seemed to be present in the greatest amounts? ____________________

____

____

(20) Would you expect similar results if leaves other than spinach had been used? ____________________

____

____

____

(21) Why are leaves green even though other pigments are present? ____________________

____

____

Summary

(1) The layer of cells that lacks chloroplasts is the ____________________.

(2) The ____________________ is the largest area of a leaf.

(3) The ____________________ layer is composed of cells that are oriented at right angles to the epidermis.

(4) The presence of ____________________ is typical of the spongy layer.

(5) The ____________________ is a waxy layer that prevents the loss of water from leaf tissues.

(6) ____________________ are composed of tissues that carry materials to and from leaf tissues.

(7) Pores found on the underside of leaves are known as ____________________.

(8) Conducting tissues in a leaf are ____________________ and ____________________.

(9) The only cells that contain chloroplasts in the epidermis are ____________________ ____.

(10) The three basic tissues of a leaf are ____________________, ____________________, and ____________________.

(11) The green color of leaves is due to the presence of the pigments ____________________ and ____________________.

(12) Red color in leaves is due to the presence of the pigment ____________________.

NAME DATE

Roots and Stems 24

MATERIALS

INVESTIGATION▶

- prepared slide, cross section of mature root (buttercup)
- woody dicot stem *(Tilia)*
- monocot stem *(Zea)*
- textbook or charts
- microscope

OBJECTIVES▶

- To become familiar with the primary growth of roots
- To study root systems and their adaptations
- To understand the structure of herbaceous and woody dicot stems
- To examine the microscopic structure of roots and stems

Problem

How is the structure of roots and stems related to their function?

When germination occurs, the seed develops into the roots and shoot. The primary root emerges first, enabling the plant to become anchored and absorb water and minerals so that the shoot can develop. Soon secondary roots begin to appear. Roots have specialized tissues that function in storing food and in conduction.

Through continued growth of the seedling, conducting tissues of the roots and stem become structurally integrated. Primary tissues develop from the meristematic regions at the tip of the stem and root. If the roots and stems of a plant live through several years of growth, they add secondary tissues, the results of the division of cells composing the cambium. The production of new cells increases the diameter of the root or stem. Monocot stems lack a cambium and therefore do not exhibit secondary growth.

Procedure

Part I. Analysis

(A) Turn to the figure of the germinating seedling on the Data Sheet. Locate the *primary root,* the main root from which *secondary roots* have begun to form. The fuzzy outgrowths are *root hairs,* which absorb water and minerals and increase the absorptive surface of the roots. Locate the *shoot* and note the relationship of the base of the shoot and the base of the root on the seedling. On the Data Sheet, label the figure of the seedling as directed. Answer questions 1–4.

(B) The primary root of many plants remains the principal root of the root system, resulting in a *taproot system.* In some plants, the root enlarges and becomes a *fleshy taproot.* In many cases, however, the primary root is short-lived and is replaced by many secondary roots. This produces a *diffuse root system.* Roots of a diffuse system may be either *thick* and *fleshy* or *slender* and *fibrous.*

Four kinds of root systems are shown on the Data Sheet. In the chart, classify the roots shown. Compare the efficiency of each system as good, fair, or poor in performing each of the activities described.

Part II. Investigation

(A) Examine the slide of a cross section of the primary root of a buttercup. The outer layer of cells is the *epidermis*. Within the epidermis are cells known as *cortex*. *Starch grains* within cortex cells indicate their storage function.

The third region to be observed is the *vascular cylinder* at the center of the root. Locate the outermost layer of thickened cells, the *endodermis*. Cells just within the endodermis are those of the *pericycle*. The cells of this layer divide to form the secondary roots.

Thickened *xylem* cells form a cross-shaped figure in the center. Located between the arms of the xylem are bundles of *phloem* cells. The *vascular cambium* occurs as a single cell layer between the xylem and phloem tissues. On the Data Sheet, label the structures: *epidermis, cortex, starch grains, endodermis, pericycle, phloem, cambium,* and *xylem*.

(B) Bark and wood are composed of specialized tissues that can only be observed with a microscope. Examine a cross section of a woody dicot stem under low power. Locate the *pith* region at the center. Notice the *vascular rays,* which radiate from the pith like irregular spokes in a wheel. Locate the *annual rings,* which indicate each season's growth. Observe that each ring is composed of two regions of cells. *Springwood* cells, formed during rapid spring growth, are found toward the inside of the ring. Smaller *summerwood* cells form during the less rapid growth periods.

Determine the age of your specimen by counting the layers of *xylem*. Xylem is composed of *vessels* and *tracheids,* specialized cells for conduction of water and minerals. Xylem *fibers* offer support to the stem. Located at the outer edge of the xylem is the *vascular cambium*. Division of cambium cells results in the production of new xylem and phloem cells. As new cells are added, the diameter of the stem increases.

Phloem cells are located just outside the cambium. Observation under high power reveals that phloem is composed of specialized food-conducting cells, the *sieve tubes* and *companion cells*. These cells, along with thick-walled phloem *fibers,* make up the inner bark.

The outer bark consists of *cork cambium* and an outer layer of thick *cork* cells. On the outline of the woody dicot stem on the Data Sheet, complete the drawing with accurate sketches of the tissues described. Label *cork, cork cambium, outer bark, phloem, inner bark, vascular cambium, xylem, springwood, summerwood, annual ring, vascular ray,* and *pith*. Answer questions 5–7 on the Data Sheet.

(C) Examine a cross section of a monocot stem under low power. Just under the outer layer of cells, the *epidermis,* are additional thick-walled cells. Together, these cells compose the *rind* of the mature stem. Examine a *fibrovascular bundle,* toward the center of the stem, under high power. Note that the bundle has the appearance of a face with distinct facial regions. Large *xylem vessels* are found in the position of the "eyes" and "nose" of the face while *phloem* occupies the "forehead." Locate *sieve tubes* and *companion cells,* which compose the phloem. The position of the "mouth" is an irregular *intercellular space*. Thick-walled *sclerenchyma cells* surround and help in support of the bundle. On the Data Sheet label the figure of the corn stem and a fibrovascular bundle. Answer questions 8–13 on the Data Sheet.

Investigations On Your Own

(1) Secondary growth of a root may be studied through the examination of a carrot or similar root. Consult your text to determine what becomes of the primary tissues as the root increases in diameter. Make drawings of the longitudinal and cross sections and label what you observe.

NAME DATE

Roots and Stems 24

DATA SHEET

Analysis

(A) On the figure of the seedling, locate and label: *primary root, secondary roots, root hairs, shoot*. Answer questions 1–5.

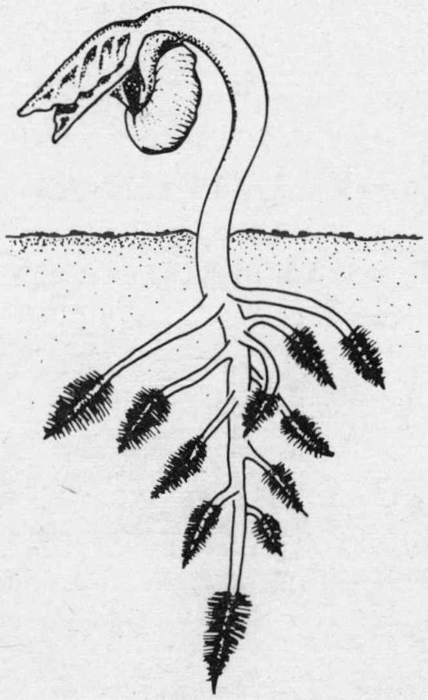

Bean seeding

(1) Describe the structure of the primary root.

(2) Where are the secondary roots developing?

(3) Locate and describe the growth of the root hairs.

(4) What function do root hairs perform?

(B) In the first space in the table on the following page, classify the roots shown below as *tap, fleshy tap, fleshy diffuse*, or *fibrous diffuse*. Compare the efficiency of each root system by rating it good, fair, or poor in performing each of the activities.

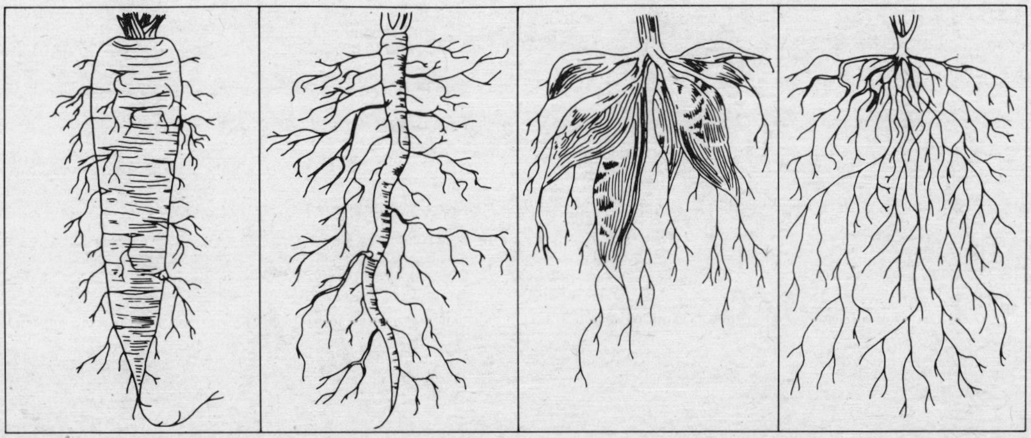

	Carrot	Dandelion	Dahlia	Grass
Type of Root System				
Anchorage				
Absorption of Minerals and Surface Water				
Absorption During Drought				
Storage				

Investigation

(A) In the space provided draw a cross section of the primary root; locate and label: *epidermis, cortex, starch grains, endodermis, pericycle, phloem, cambium, xylem*.

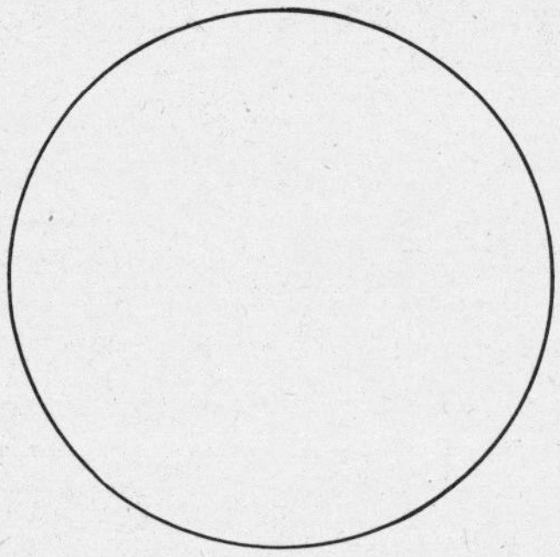

(B) Within the pie-shaped outline, draw in the cellular detail of a woody dicot stem. Pay close attention to the description given in the Procedure. The fourth line from the center represents the vascular cambium. You may need your textbook to locate the cells and tissues as described. When the drawing is completed, label the figure. Brackets may be used where necessary. Label: *cork, cork cambium, outer bark, phloem cells, inner bark, vascular cambium, xylem, springwood, summerwood, annual ring, vascular ray, pith*.

NAME DATE

(5) What are some functions of the cork?

(6) What is the function of the cork cambium?

(7) What tissues compose the inner bark?

What evidence is there that the stem has lived for more than a single year?

Explain why the annual rings are of unequal thickness.

(C) On the general view of the corn stem, label: *epidermis, rind, pith, fibrovascular bundle*. On the figure of the fibrovascular bundle label: *xylem, phloem, intercellular space, pith, sclerenchyma*.

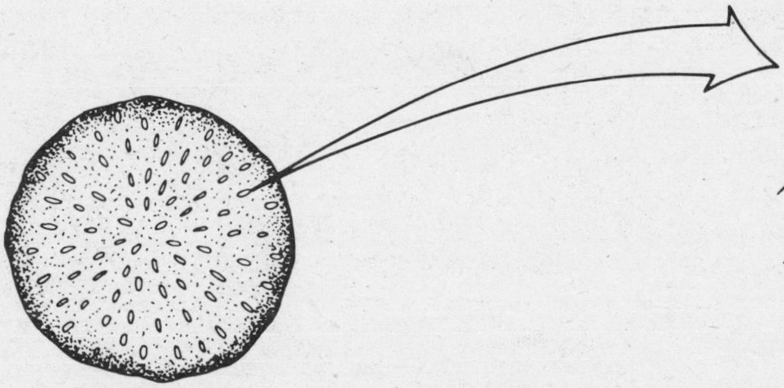

Monocot stem (cross section)

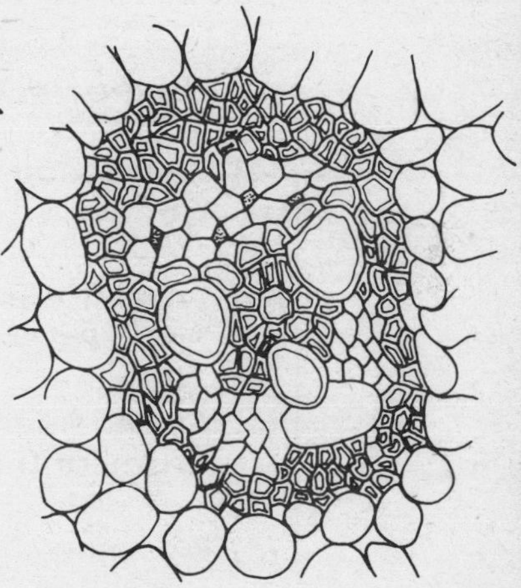

Fibrovascular bundle

(8) Describe the appearance of the epidermal cells.

(9) What tissue occupies most of the stem?

(10) Describe the cells that compose this tissue.

(11) Where in the stem are the fibrovascular bundles most numerous?

(12) What effect does the arrangement of cambium have on a monocot stem? ________

____

____

(13) What might be the function of the thick-walled sclerenchyma fibers that surround a bundle? ____

Summary

(1) Explain how the maturation of certain root tissues and the continual addition of vascular tissues to a root contributes to the over-all functioning of a plant.

____

____

____

____

____

____

(2) In the chart below, summarize the similarities and differences between the monocot and dicot stems you have observed and studied.

Characteristic	Dicot	Monocot
(a) Herbaceous or Woody		
(b) Type of Vascular Tissue Present		
(c) Arrangement of Vascular Tissue		
(d) Presence of Vascular Cambium		
(e) Location of Pith		
(f) Means of Support		
(g) Probable Life Duration of Stem		

NAME DATE

25 Water Relations

MATERIALS

INVESTIGATION▶

- fresh leaves of: Tulip, *Tradescantia*, Geranium, or *Kalanchoe*
- healthy geranium plant
- slide, cover glass
- microscope
- forceps
- paper toweling
- 10% salt solution
- 2 sandwich bags
- cobalt paper
- paper clips
- cellophane tape

OBJECTIVES▶

- To study the pathway of water through a plant
- To observe the functioning of stomata
- To demonstrate water loss through transpiration

Problem

How does water move through a plant?

An abundant supply of water is necessary for the cells of plants to remain alive and functional. Roots absorb water from the soil through the root hairs. After passing through the cells of the cortex, water molecules become part of the continuous column of water of the xylem cells of the central cylinder. This water is conducted up from the roots through the xylem cells of the stem and into the smallest veins of the leaf. From the leaf veins, water is given up to the mesophyll cells of the leaf for use in photosynthesis and other plant processes.

The amount of water actually used in plant activities is only a fraction of that absorbed by the roots. Ninety-nine percent of the water absorbed by the roots escapes through the stomata of the leaves in the process called transpiration. The movement of water through a plant is a remarkable process. Conduction is accomplished by a combination of forces including root pressure, capillarity, and transpiration.

Procedure

Part I. Analysis

To understand how water moves through a plant it is necessary to understand how each organ functions in conduction. Study the figure of the stylized plant on the Data Sheet. Determine and indicate with small arrows the pathway of water from the soil into the roots and its conduction upward to the leaves. Label the figure with the plant tissues and structures involved in water movement. You may want to refer to your text as you answer questions 1–11.

Part II. Investigation

(A) In addition to providing for the movement of gases in and out of a leaf, the stomata have the function of regulating movement of water through the plant.

Remove a small piece of the lower epidermis from a tulip, geranium, or *Kalanchoe* leaf by tearing the blade, twisting as you tear. The epidermis appears as a thin transparent skin. Remove a portion of the epidermis with forceps.

Prepare a wet mount of a piece of the epidermis, being careful not to wrinkle the tissue. Make sure that the outer surface of the epidermis is up when you place the tissue on the slide. Examine under low power. Note the shape and arrangement of the *epidermal cells* and the number and arrangement of the *stomata* (sing. stoma). Locate the bean-shaped *guard cells* on either side of a stoma. Closely examine the walls of the guard cells. The guard cells regulate the size of the stoma by changing shape.

Turn to the Data Sheet and, in the space provided, draw several epidermal cells with guard cells surrounding a stoma. Answer questions 12–15.

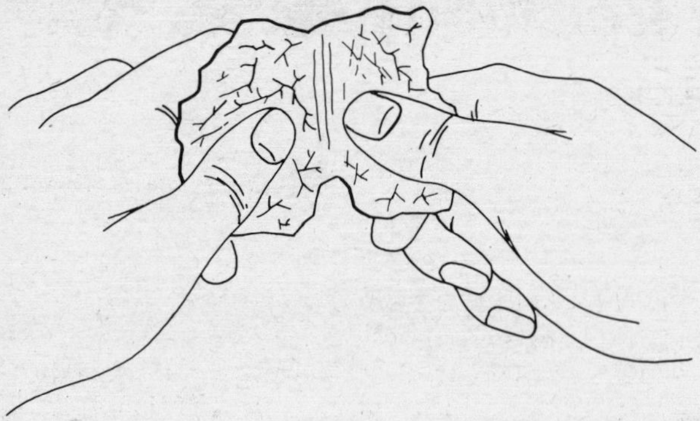

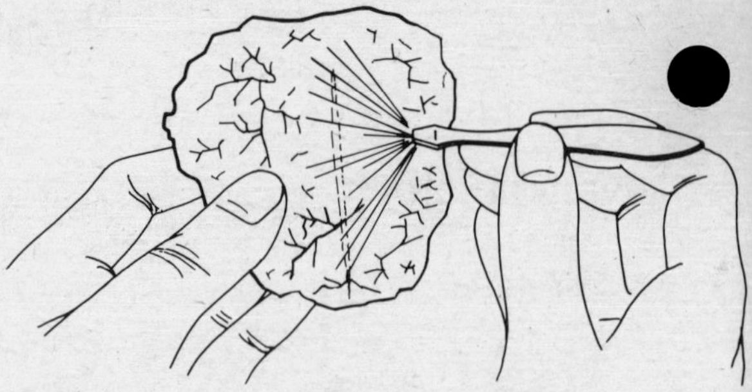

(B) To observe the opening and closing of the stoma, prepare a fresh, wet mount of the lower epidermis of a leaf exposed to bright light. Examine under low power. Determine if the stomata are open or closed. Remove the water from the preparation by placing paper toweling at the left edge of the cover glass to draw out the water. Add a 10% salt solution to the right side of the cover glass. Observe quickly to notice what happens to the guard cells and stoma. Turn to the Data Sheet and answer questions 16–19.

(C) Transpiration is essentially the loss of water from a plant through evaporation from leaf tissues. Remove a leaf from a vigorously growing geranium plant. Fasten a strip of cobalt paper to the lower and upper surface by clipping the paper at both ends of the leaf as indicated in the figure. Do not damage the leaf unnecessarily. Blue cobalt paper turns pink in the presence of moisture.

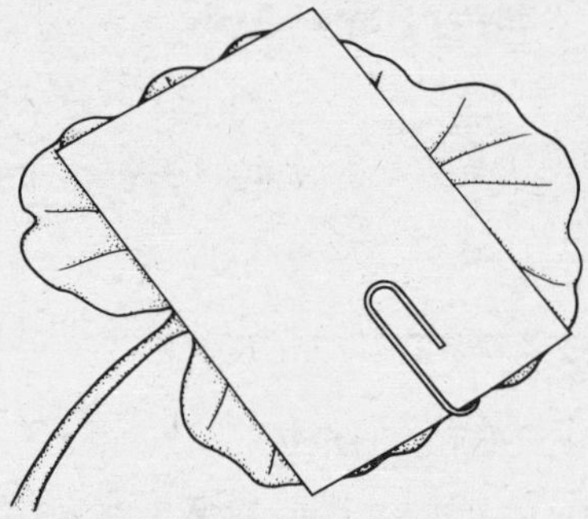

Place the leaf in a small plastic sandwich bag and seal it with tape. Place a strip of cobalt paper in a bag by itself and seal it. Observe the leaf after 10–15 minutes. After observation, turn to the Data Sheet and answer questions 20–29.

Investigations On Your Own

(1) In the examination of the structure of roots and stems, it may be observed that certain tissues are responsible for the conduction of substances up and down the stem. To determine what tissues are involved with water conduction, fill a 250-mL flask about one-third full of water colored with powdered eosin or red ink. Place a freshly cut stem of coleus in the flask. After at least 30 minutes, remove the stem and split it lengthwise with a razor blade. Record your observations in a drawing. Give an explanation for your conclusions.

(2) Devise an experiment to demonstrate how much water passes through a plant for a given period of time. Use a potted plant and weigh it at intervals. Determine accurately measurements of initial weights and succeeding weights of the plant. Record your observations in a graph.

NAME DATE

25 Water Relations

DATA SHEET

Analysis

On the diagram of the plant, indicate with small arrows the pathway of water from the point of absorption in the roots to the point where transpiration takes place in the leaves. On the figure, label the following areas or structures: *soil surface, root hairs, root epidermis, cortex, xylem of root, xylem of stem, leaf petiole, midrib of leaf, mesophyll, air space, stoma, guard cells, atmosphere*. Answer questions 1–11.

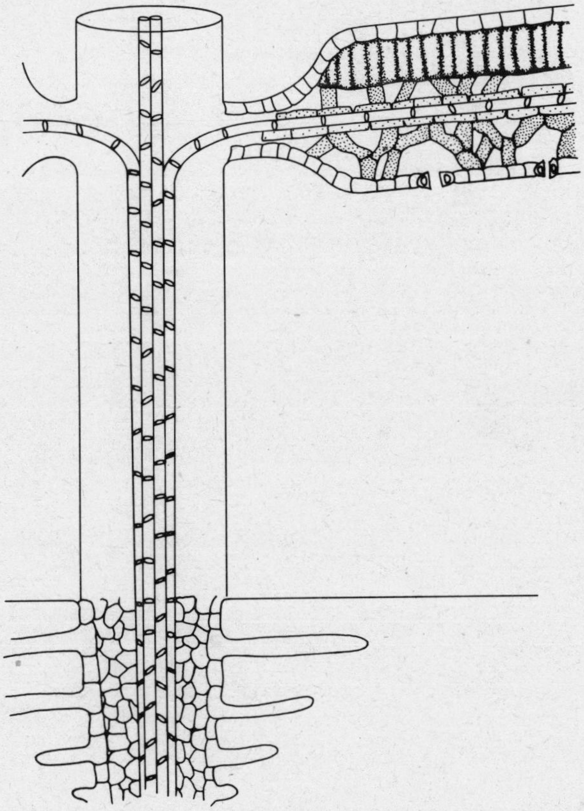

(1) In what structure is water absorbed by a root? ______________________

____

(2) What mechanism is involved in the absorption of water? ______________________

____

(3) What tissues of the root carry water upward to the stem? ______________________

____

(4) In what tissues of the stem and leaf is water conducted? ______

______

(5) What is meant by a "column of water" as it exists in a plant? ______

______

______

______

______

(6) What becomes of the water when it reaches the mesophyll cells of the leaf? ______

______

______

(7) What becomes of the water in the intercellular spaces? ______

______

(8) What is this process called? ______

(9) In terms of water molecules, what is cohesion? ______

______

(10) How does the "transpiration-cohesion" theory account for the movement of water through the plant? ______

______

______

______

______

______

(11) What other factors may contribute to water movement? ______

______

Investigation

(A) In the space provided, draw a portion of the leaf epidermis with several stomata. Label: *epidermal cells, guard cells, chloroplasts, stomata*. Answer questions 12–15.

(12) Are chloroplasts present in the epidermal cells? ______

(13) What variation in thickness can be observed in the inner and outer walls of the guard cells? ______

______

(14) What structures can be observed in the guard cells? ______

______

(15) Why is it important that the guard cells change shape? ______

______

(B) Prepare a fresh specimen of epidermis and examine under low power. Answer questions 16–19.

(16) Are the stomata open? ______

______

(17) What happened to the guard cells and the stoma after adding the salt solution?

______

______

(18) How can you explain the changes that occurred? ______

______

______

______

(19) What natural activity of a leaf might cause the stomata to open? ______

______

______

______

(C) Follow the directions in the procedure for demonstrating transpiration. Answer questions 20–29.

(20) What is the purpose of the second bag in which cobalt paper was placed?

(21) After fifteen minutes, what was the color of the cobalt paper on the upper surface?

(22) On the lower surface?

(23) What is the color of the cobalt paper in the empty bag?

(24) How do you account for the difference in the color of the cobalt paper?

(25) Why is it necessary to seal the bags?

(26) Are the leaves wet?

(27) From what part of the leaf did the moisture come?

(28) Why is a control bag needed?

(29) Originally, how did the water get to the leaf?

NAME DATE

Summary

(1) Summarize the way in which the structure of the leaf enables it to function ideally in the movement of water through the plant body.

(2) On the basis of what you have learned about transpiration, account for the fact that leaves often wilt on a hot summer day but return to normal at night.

(3) List as many factors as you can think of that might affect the loss of water from a plant.

(4) The following events trace the movement of water from the soil, through a plant, and into the atmosphere. Using your knowledge of plant structure, unscramble the pathway and write the events in order of occurrence in the spaces provided. The first and last are done for you.

Water

4. enters the petiole of the leaf	**(a)** water in the soil
5. moves across the cortex of the root	**(b)** ______________________
6. absorbed by mesophyll cells	**(c)** ______________________
7. enters vascular cylinder	**(d)** ______________________
8. enters veins of leaf	**(e)** ______________________
9. absorbed by root hairs	**(f)** ______________________
10. leaves leaf through stomata	**(g)** ______________________
11. moves into xylem cells of stem	**(h)** ______________________
12. evaporates to air spaces of leaf	**(i)** ______________________
13. moves up xylem of stem	**(j)** ______________________
	(k) ______________________
	(l) water escapes to atmosphere

NAME DATE

Plant Growth and Responses 26

MATERIALS

INVESTIGATION▶

- 2 bean seedlings grown with light on one side
- 2 bean seedlings grown with light overhead
- 2 bean seedlings grown in darkness
- rubber bands
- blotter paper
- 2 Petri dishes
- corn grains
- bean seeds
- half-gallon milk container

OBJECTIVES▶

- To study how auxins affect plant growth
- To observe the effect of light on plant growth
- To observe the effect of gravity on plant growth

Problem

What factors are responsible for growth and development in plants?
Biologists have found that in addition to environmental factors, growth in plants is regulated by chemical substances called hormones. Plant hormones, although produced in small amounts, have a great influence on plant growth. Hormones are transported from the region in which they are produced to other plant organs where they influence growth and other physiological activities. Thus, growth in one organ of a plant may be dependent upon the cellular activities of another.

Growth responses of plants are called tropisms. Tropisms may be toward or away from the stimulus and are thus designated as being positive if toward or negative if away. Plants exhibit tropic responses to environmental stimuli such as light, gravity, and touch.

Procedure

Part I. Analysis

(A) Auxins produced in the terminal bud influence the growth patterns of plants by inhibiting the growth of lateral buds. To better understand this influence, turn to the Data Sheet and study the figures of the growth patterns of two different coleus plants. As you study the plants, give thought to the auxin activity that has occurred in each plant. Plant A is the control plant while plant B is the experimental plant. After reaching some conclusions, you may wish to consult your textbook as you answer questions 1–6 on the Data Sheet.

Part II. Investigation

(A) Obtain two bean seedlings that have been grown under each of the following conditions:

(1) light received only from one side
(2) darkness
(3) light received from overhead

Observe the seedlings and interpret their growth on the basis of auxin influence. You may want to use your textbook as a guide while answering the questions. On the Data Sheet answer questions 7–12.

Exchange the plants that grew directly under light with those illuminated from one side. Observe the plants periodically for several days and note the changes that occur in the growth patterns of the leaves and stems. Record your observations in the chart on the Data Sheet.

(B) *Geotropism* is the growth response of a plant to gravity. This tropic response can be observed by preparing the following demonstration: Cut a piece of blotter paper to fit snugly into the top of a Petri dish. Place three beans on a line across the center as indicated in the figure. Now, place the bottom of the Petri dish so that it pushes against the beans and holds them in place. Use a small amount of tape at the sides to hold the two sides securely together. Prepare another dish in the same manner with three corn grains. Place both Petri dishes in the cut lower-quarter of a half-gallon milk carton filled one-third full of water. Use rubber bands going around the carton on the front and back sides to hold each dish upright. Allow germination to occur.

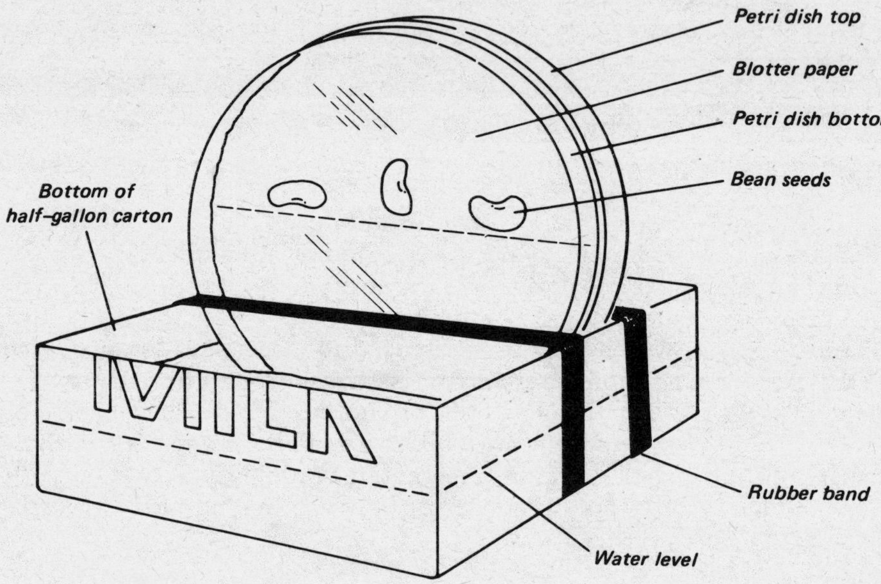

Observe the direction of root and shoot growth from each of the seeds. After drawing conclusions from your observations, answer questions 13–16 on the Data Sheet.

Investigations On Your Own

(1) To observe phototropic responses further, examine the arrangement of the leaves on plants. See if you can locate a plant in which the blade of a leaf completely shades the blade of another. Leaf orientation allowing each blade to receive a maximum amount of light and a minimum amount of shade is known as leaf mosaic. Write up your findings of the plants observed.

(2) Thigmotropism is the response on the part of plants to the stimulus of contact or touch with another object. Observe the twining of morning glory vine, bindweed, or garden peas. Carefully remove the plants from their support, observing the direction of twining (clockwise or counterclockwise). Observe daily and record the progress of the plant in gaining new support. Write your daily observations, including the direction of twining, and growth responses of the plant since removal from its support.

NAME DATE

Plant Growth and Responses 26

DATA SHEET

Analysis

Study the figures of the two coleus plants. Note differences in their structure and branching patterns. On the figures, label: *terminal bud, lateral buds, main stem, lateral stems*.

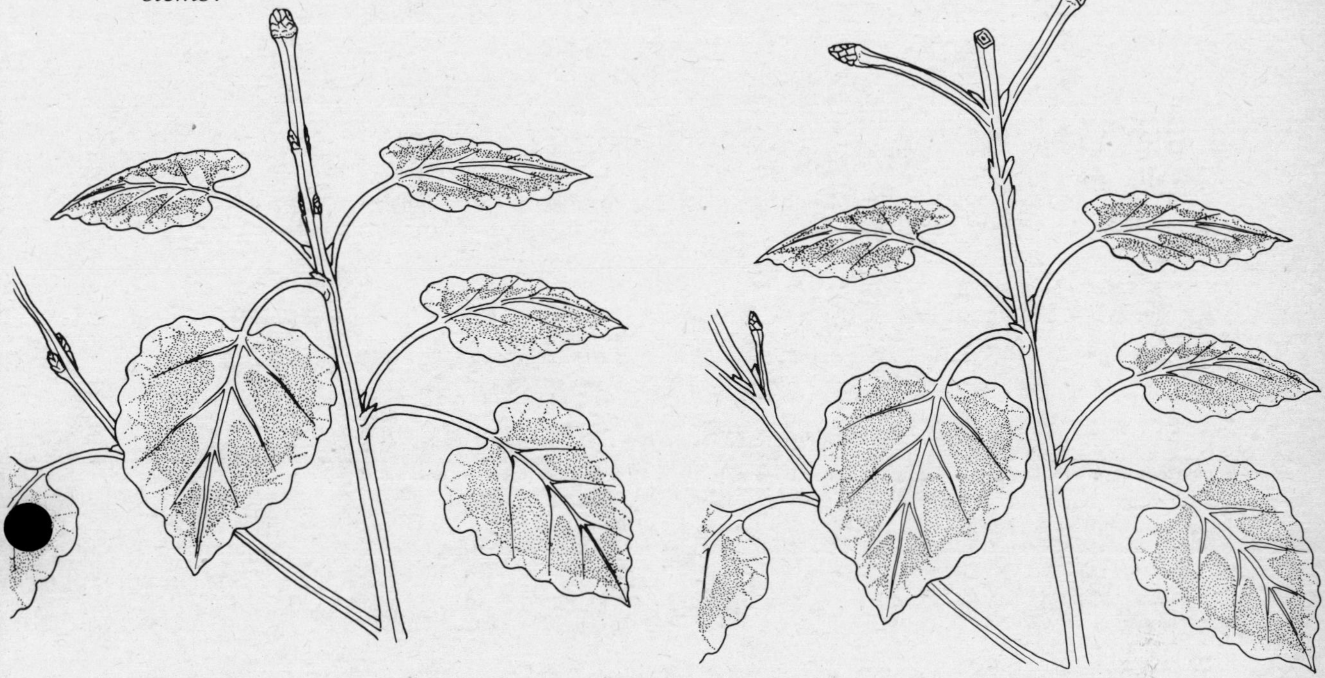

(1) Where is the terminal bud located? ______

(2) The lateral buds? ______

(3) What becomes of the auxin produced in the terminal bud? ______

(4) What is the effect of the auxin on the lateral buds? ______

(5) What differences can be noted between Plant A and Plant B? ______

(6) On the basis of your conclusions, explain what is meant by *apical dominance*.

Investigation

(A) Examine the seedlings.

(7) How have the seedlings responded to light received from one side?

(8) How are the leaves oriented?

(9) Describe the growth of the stem.

(10) Describe the growth of the seedlings receiving light from overhead.

(11) How does the growth of the seedlings grown in the dark compare with that of those grown in the presence of light?

(12) Consult your textbook and explain the growth response of each type of seedling on the basis of auxin content.

Light from one side:

NAME DATE

Light directly overhead: ______

Darkness: ______

Follow the directions given in the procedure. Record your observations in the chart below.

Day	Date	Changes in Stem	Changes in Leaves
1			
2			
3			
4			
5			
6			
7			
8			

(B) Geotropism

(13) Describe any similarities and differences in the germination of the beans and corn grains. ______

(14) Does the position of the seed have any effect on the growth of the root and shoot? ______ Account for your answer. ______

(15) Do roots show a positive or negative response to gravity? ______________

(16) The shoot? ______________

Summary

(1) Discuss how the elimination of apical dominance is the basis for the pruning of hedges and other plants. ______________

(2) What is a tropism? ______________

(3) How does unequal growth cause bending of stems? ______________

(4) When a potted plant is on its side, the stem curves upward away from gravity. Explain why most of the curvature occurs near the tip. ______________

(5) In a natural situation, what value do tropisms have for plants? ______________

NAME DATE

Plant Reproduction 27

MATERIALS

INVESTIGATION▶

- flowers such as sweet pea, lily, sunflower, snapdragon, willow, *Avena*
- single-edged razor blade
- microscope
- prepared slide of cross section of lily anther
- slide
- hand lens or stereoscopic microscope
- dropper
- forceps

OBJECTIVES▶

- To study the structure of dicot and monocot seeds
- To understand methods of fruit and seed dispersal
- To identify the parts of the flower
- To understand the reproductive function of flower parts

Problem

How do flowering plants accomplish reproduction?

Flowering plants are one of two large groups of plants that form seeds to reproduce themselves. One basis for classifying flowering plants is the number of cotyledons or seed leaves within the seed. Whether there be one or two cotyledons, the seed is a matured ovule and is the final product of angiosperm reproduction.

The new plant within the seed is provided with stored food and special coverings. The seeds of some plants are consumed by man and many others are planted to grow into consumable products. In natural situations, animals, wind, and water are the agents by which seeds are dispersed. Many seeds are adapted for dispersal by these agents. Before there can be seeds, however, there must be a flower. All seeds, however large or small, are the products of flowers.

Flowers vary from one plant to another and their structure is another basis for classification. Some plants have their flower parts reduced to those essential for seed production, the stamen and the pistil. Those flowers with which we are most familiar also have showy sepals and petals. These are accessory parts and may have only a passive role in seed formation.

Procedure

Part I. Analysis

(A) The lima bean is an example of a dicot seed. Turn to the Data Sheet and study the figures of the external and internal view of the lima bean. Locate the scarlike *hilum* on the concave side. Above the hilum may be found the *micropyle.* The outer seed coat is the *testa.* When the testa is removed, the seed divides into two parts. This reveals the embryo plant, which remains attached to one of the two *cotyledons.* Note the *plumule,* consisting of two tiny leaves that enclose the *terminal bud* of the future plant. Below the plumule is the *hypocotyl,* the embryonic stem. Locate the *radicle* at the base of the hypocotyl. The radicle is the embryonic root. On the Data Sheet label the external and internal views of the bean. Answer questions 1–5.

(B) The corn grain is an example of a monocot seed. Examine the figure of the corn grain. At the base of the grain is a pointed projection, the *point of attachment* to the cob. Another projection at the top of the grain is called the *silk scar.* At this point, the silk of the style was attached to the grain during the flowering stage. Study the figure of the longitudinal section of the corn grain. The outer covering is the *ovary wall.* The lower portion contains the embryo and *cotyledons.* The upper part of the embryo is the plumule and the surrounding *epicotyl sheath.* The bulk of the grain is *endosperm* tissue that supplies food to the embryo. On the Data Sheet label the external and internal views of a corn grain. Answer questions 6–10.

(C) Fruits and the seeds contained within are dispersed by a variety of methods. Some plants are so constructed that seed dispersal is accomplished by the parent plant itself. In most cases, however, dispersal is dependent on some other means, such as animals, wind, or water. Examine the drawings of seeds and fruits on the Data Sheet. Use your text to help you identify each fruit or seed as *fleshy* or *dry.* Determine the type of fruit or seed each is. Speculate on how each may be dispersed. Complete each column on the chart found on the Data Sheet.

Part II. Investigation

(A) Examine your flower specimen and locate the parts that compose a complete flower as described in your textbook. Parts essential for seed production are the stamen and pistil. Nonessential parts are the petals, sepals, and supporting structures. On the Data Sheet label the parts. Answer questions 21–28.

Remove a stamen with forceps and examine it using a hand lens. Prepare a wet mount of some pollen grains by dusting the anther on a slide. Examine the pollen grains under low and high power of the microscope. Examine a prepared slide of the cross section of a lily anther under low power. The pollen grains contain the male sex cells produced as a result of meiosis from special cells within the anther. Turn to the Data Sheet and label the cross section of the anther. Answer questions 29–33.

Carefully remove the pistil from the flower and examine it closely under a stereoscopic microscope. Using a sharp, single-edged razor blade, make a wet mount of a cross section of the ovary. Examine it under a stereoscopic microscope. Notice that the ovary is divided into sections known as *carpels.* Each carpel contains serveral *ovules.* The ovules extend into a cavity known as a *locule.* While eggs are not visible, each ovule contains an egg produced by meiosis from special cells within the ovule. Observe that each ovule is attached to the ovary wall through a tiny stalk, the *placenta.* Turn to the Data Sheet and label the cross section of the ovary. Answer questions 34–37.

(B) Examine the flowers provided by your teacher. Flowers have several features used in their classification. *Regular* flowers are symmetrical, *irregular* are not. Some flowers have several small flowers in one receptacle. They are called *composite* flowers. *Staminate* flowers are those with stamens only. *Pistillate* flowers have only a pistil present. In some flowers the sepals are *petaloid.* Study each flower closely and describe the flowers, using the italicized terms, above, that are applicable. Place your descriptions in the space provided on the Data Sheet.

Investigations On Your Own

(1) Germination of the pollen tube can be observed by transferring pollen grains to a drop of 10% sucrose solution and making a hanging drop slide. Ask your teacher to demonstrate how to make such a preparation. Use several different kinds of pollen and take note of the time required for germination to occur. You may also want to vary the concentration of the sucrose solution. Write up your observations in a report.
(2) Prepare a germination jar according to instructions from your teacher. Use bean seeds and corn grains to observe differences and similarities in the germination of dicot and monocot seeds. Make daily observations and summarize your observations in a series of accurate sketches. Indicate what becomes of the various seed structures as growth of the seedling progresses.

NAME ______________________________ DATE ______________

Plant Reproduction 27

DATA SHEET

Analysis

(A) Study the figures of the external and internal view of the bean seed. On the figure of the external view of the bean, label: *hilum, micropyle, testa*. On the figure of the internal view, label: *cotyledons, plumule, hypocotyl, radicle, testa*.

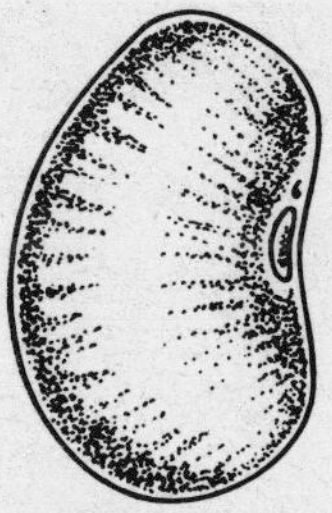

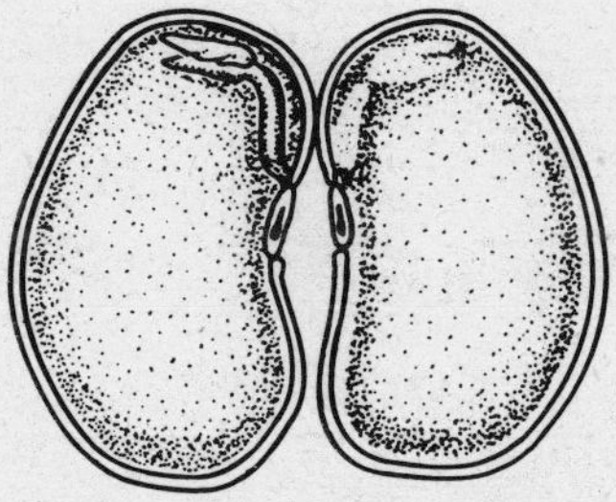

(1) What does the hilum represent? ______________________________

(2) What is the significance of the micropyle? ______________________________

(3) Would you expect all seeds to have a hilum and a micropyle? ______________

Explain your answer. ______________________________

(4) What is the function of the cotyledons? ______________________________

(5) What becomes of the food stored in the cotyledons during germination? __________

(B) Study the figures of the external and internal view of the corn grain. On the figure of the external view, label: *point of attachment, silk scar*. On the figure of the internal view, label: *silk scar, endosperm, cotyledon, plumule, hypocotyl, radicle, point of attachment*.

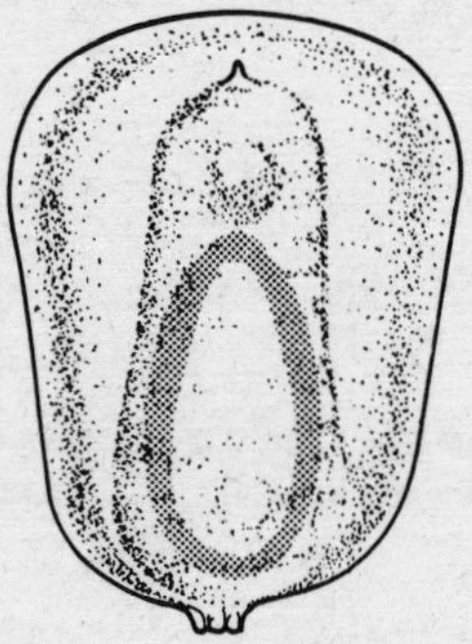

External

Cross section

(6) Explain why an ear of corn is the product of a group of flowers. ____________________

____

____

____

(7) Account for the location of the silk scar. ____________________

____

____

(8) If an ear of corn had 250 grains, how many silks would there have been? Explain your answer. ____________________

____

____

____

(9) Would you expect to find a hilum and micropyle in the corn grain? Explain what you are able to locate. ____________________

____

____

____

(10) Where is food stored in the corn grain? ____________________

(C) Study the fruits and seeds shown in the figures on the next page. Fill in the chart, giving the classification of the fruit as *fleshy* or *dry*. If the fruit is dry and splits open when ripe, the fruit is *dehiscent*. If splitting does not occur, the fruit is *indehiscent*. With the aid of your text, determine the type of fruit such as pome, berry, achene, etc. Determine the methods by which the fruit or the seeds within are dispersed under natural conditions.

NAME DATE

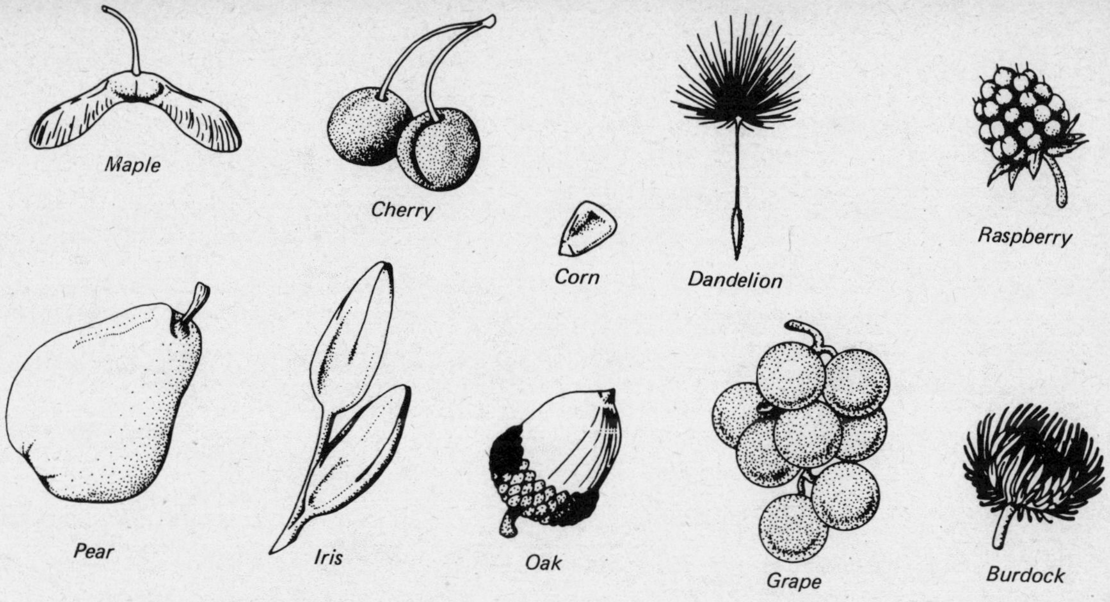

Name of Fruit	Classification	Type	Method of Dispersal
(11) Maple			
(12) Cherry			
(13) Dandelion			
(14) Raspberry			
(15) Pear			
(16) Iris			
(17) Oak			
(18) Grape			
(19) Burdock			
(20) Corn			

Investigation

(A) On your flower specimen, locate the parts of a flower as discussed in your textbook. On the figure label: *pedicel, receptacle, sepals (calyx), petals (corolla), anther, filament, stamen, stigma, style, ovary, pistil.*

(21) Describe the number and color of the sepals. ____________________

____

(22) Describe the number and appearance of the petals. ____________________

____

(23) Why are the stamen and pistil known as essential parts? ______________________

____

____

(24) How does the number of stamens compare with the parts already observed? ______

____

____

(25) What seems to be the number plan of the flower? ______________________

____

(26) Of what group of flowering plants is this characteristic? ______________________

____

____

(27) Why is it necessary that the stigma be sticky? ______________________

____

(28) What is produced within the ovary? ______________________

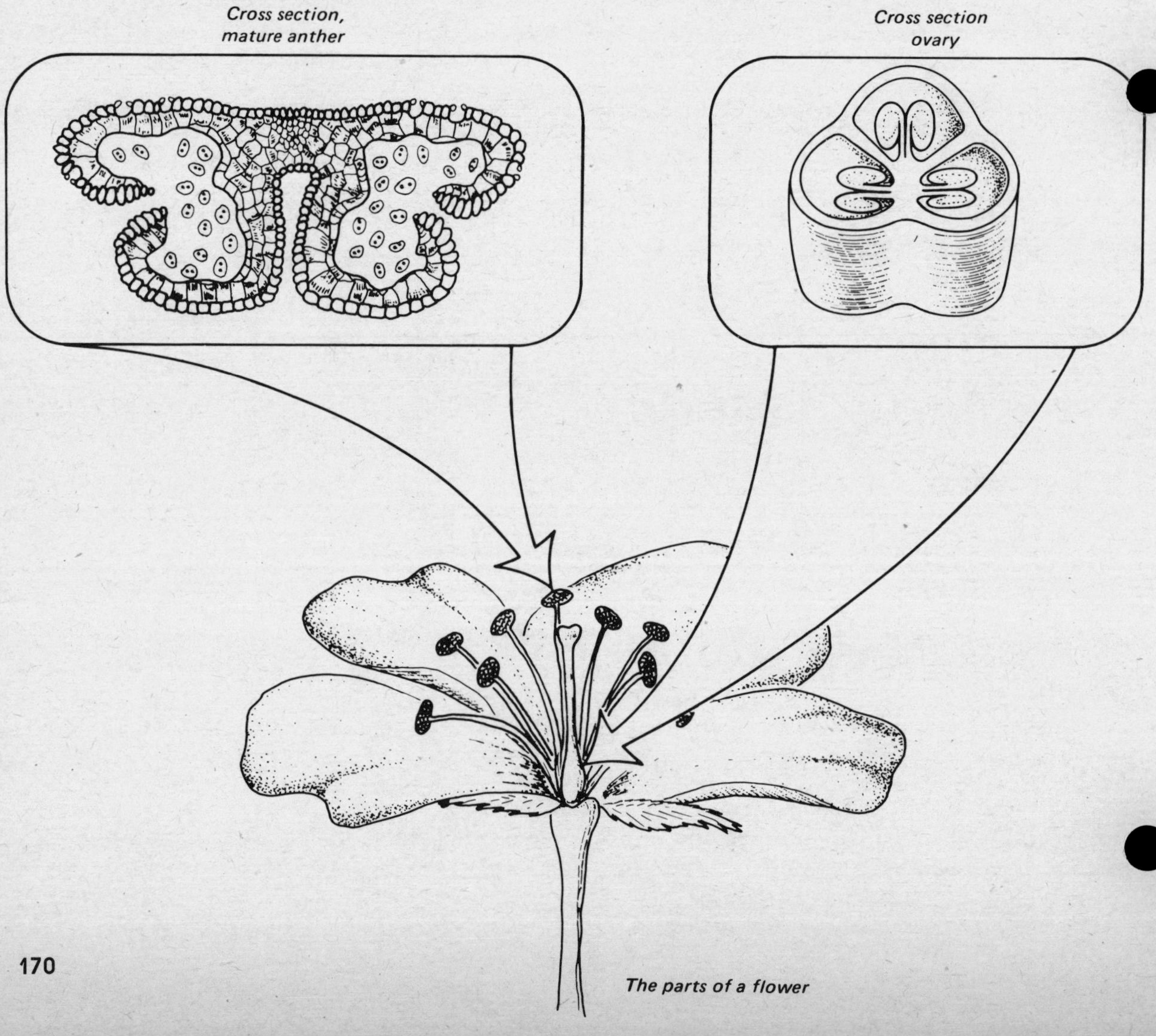

The parts of a flower

(B) Follow the directions as outlined in the Procedure. After examination of the microscopic cross section of the lily anther, label on the figure above: *pollen sac, pollen grains*.

(29) Describe what you observe in the examination of the anther. ______

______

______

______

(30) Describe the surface of the pollen grains. ______

______

(31) How many cavities or pollen sacs are seen? ______

(32) When the anther matures, how is pollen released? ______

______

______

(33) What becomes of the pollen as it is released? ______

______

______

(C) Follow the directions as outlined in the Procedure. On the figure of the flower, label the cross section of the lily ovary. Label: *ovary wall, carpel, ovule, locule*.

(34) In relation to pollination, suggest a reason for the stigma being supported as it is by the style. ______

______

(35) How many cavities or carpels make up the ovary? ______

______

(36) Explain the significance of the number. ______

______

(37) Why must the ovules be attached as they are? ______

______

______

______

(B) Study each flower closely to determine which of the flower descriptions below are applicable. On the chart, describe the flower using the following terms: *regular, irregular, composite, staminate, pistillate, petaloid* sepals.

Name of Flower	Description of Flower

Summary

(1) In the following chart, briefly relate how each flower part contributes to the function of reproduction.

Floral Part	Function
sepal	
petal	
filament	
anther	
stigma	
style	
ovary	

NAME DATE

The Hydra – A Coelenterate 28

MATERIALS

INVESTIGATION▶

- hydra culture
- *Daphnia* culture
- dissecting microscope
- Syracuse watch glass
- dropping pipette
- probe
- flashlight (penlight type)
- piece of blue cellophane
- 1% acetic acid

OBJECTIVES▶

- To observe the structure of a living coelenterate
- To observe behavior of a living hydra

Problem

What coelenterate characteristics can be observed in a living hydra?

Most coelenterates live in the marine environment. Some have the form of a *medusa* and some, of a *polyp*. Some have both forms during their life history.

The hydra is a simple coelenterate found in freshwater ponds and slow-moving streams. Hydras may be collected and kept in the laboratory for observation. This tiny polyp attaches itself to plants and debris in the water by its basal disk. It may release its attachment and move by a somersaulting motion or float to a new location. The animal catches small organisms with its tentacles, paralyzes them with stings, and then takes the prey into its digestive cavity. These are very interesting organisms to observe.

Procedure

Part I. Analysis

Turn to the Data Sheet. Compare the medusa and the polyp forms of coelenterates. Then, identify the various coelenterate characteristics of the hydra. Refer to your text to help you. Answer questions 1–3 on the Data Sheet.

Part II. Investigation

Using a pipette, carefully transfer a hydra to a Syracuse watch glass. Add enough water to cover the animal. Watch the hydra with the aid of a dissecting microscope using diffused light. Observe the basal disk, by which the animal attaches itself to a surface. Notice how it moves and perhaps changes its shape.

The tentacles are characteristic structures of coelenterates. The tentacles surround a small slit, the mouth. Answer questions 4–6 on the Data Sheet after you have made these observations.

While observing the hydra, gently tap the rim of the Syracuse watch glass with a pencil. Describe your observations in question 7 of the Data Sheet. Now, use the probe and gently touch the animal in various areas, such as the tentacles, mouth, body, and basal disk. Record your observations in question 8 of the Data Sheet and answer questions 9 and 10.

As you are observing the hydra with the dissecting microscope, have your lab partner put *one* drop of 1% acetic acid at one side of the Syracuse watch glass. Observe the tentacles as the acid diffuses through the water. Rinse out the Syracuse watch glass and put in fresh water. Place several hydras in the glass. Shine a light on one side and observe any reaction. Now, shine the light through the blue cellophane on one side of the watch glass. Record your observations in questions 11–13 on the Data Sheet.

Select a few small *Daphnia* and use a pipette to transfer them to the watch glass containing the hydras. Observe the reaction of the hydras to the *Daphnia*. Answer questions 14–16 on the Data Sheet.

Investigations On Your Own

(1) Hydras are able to regenerate themselves from pieces. You may wish to design and carry out an experiment to show this.

(2) *Aurelia* belongs to the same class of coelenterates as the large jellyfish. It is found in marine waters all over the world. The medusa stage in the life cycle is the most obvious. If you have a preserved specimen available, examine it carefully. Make a diagram and label the major structures. Compare the medusa form of *Aurelia* to the polyp form of a hydra.

(3) Feed some *Daphnia* to hydra as you did in this Investigation. After the prey has become quiet, remove one and examine it with a compound microscope. Using charts and reference books, identify the nematocysts and make drawings of your observations.

NAME ________________________________ DATE ________

The Hydra – A Coelenterate 28

DATA SHEET

Analysis

Compare the polyp and medusa forms of the body plan of a coelenterate. Label each diagram as *polyp* or *medusa*. Then, label the hydra: *mouth, tentacle, epidermis, gastrovascular cavity, nematocyst, basal disk*.

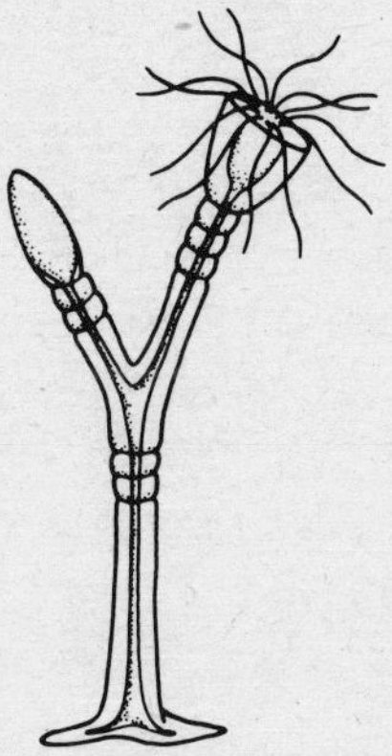

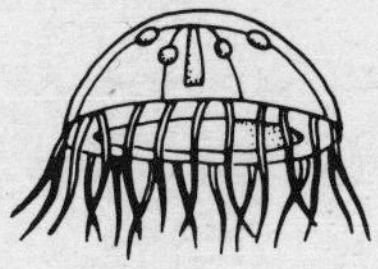

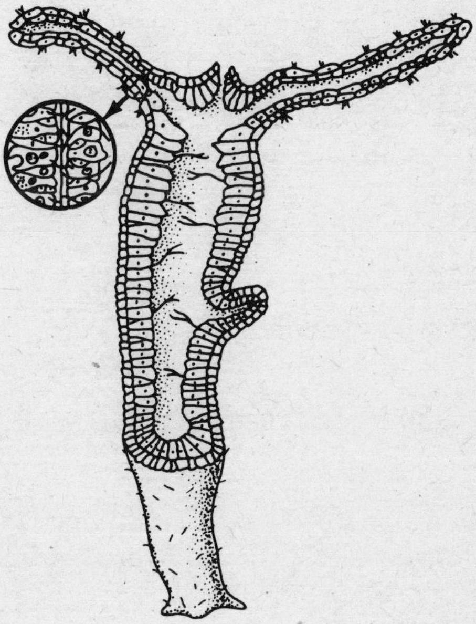

Analysis

(1) What body form does a hydra represent? ________________________________

(2) What characteristics of the hydra are present in the polyp? ________________

(3) What characteristics of the hydra are present in the medusa? ________________

____

Investigation

(4) How many tentacles are on your hydra specimen? ________________________

(5) Do all the hydra have the same number of tentacles? ____________________

(6) Describe any movements you have observed. ___________________________

____

(7) Describe the reaction of the hydra to the stimulus of tapping the watch glass with a pencil. ____

(8) Record the behavior of the hydra to the stimulus of touch. ______

(9) Did you observe anything to indicate that touch stimulus spread to areas distant from the contact? ______

(10) How can you explain this? ______

(11) Record the reaction of the hydra to acetic acid. ______

(12) Record the reaction of the hydra to light. ______

(13) Record the reaction of the hydras to blue light. ______

(14) Describe how a hydra captures *Daphnia*. ______

(15) What evidence indicates the use of stinging cells by the hydra? ______

(16) How does the hydra ingest its food? ______

Summary

What observations in the behavior of a hydra indicate coordination? ______

NAME DATE

Adaptations of Worms 29

MATERIALS

INVESTIGATION▶

- living planarians
- pipettes
- Syracuse watch glass
- hand lens (or dissecting microscope)
- penlight flashlight
- dissecting needles
- liver
- glass slide
- acetic acid
- paper toweling
- probe
- red light source
- loose, moist soil
- live earthworm
- preserved earthworm
- charts of earthworm anatomy
- dissecting pan
- scissors
- scalpel
- straight pins
- charts of tapeworm
- preserved specimen of tapeworm

OBJECTIVES▶

- To observe and compare the behavior of planarians and earthworms
- To examine the external adaptations of an earthworm
- To identify internal structures of an earthworm

Problem

What adaptations can be observed in free-living worms?

Although many worms are parasites and cause illness in their hosts, a great many worms are free-living. The planarians may be collected in a slow-moving stream or pond. Examine the undersides of stones and sticks carefully. These animals look like small droplets of gray, brown, or white jelly. They can also be collected by tying string to a piece of liver and letting the liver remain on the bottom of a quiet stream or pond for a few hours. When the planaria are transferred to the collecting jar, they will begin to crawl along the sides and bottom.

Perhaps the earthworm or nightcrawler is a more familiar worm. Living in the ground, it literally eats its way through the soil, providing tiny tunnels for air and water. Its wastes further enrich the soil for plants. During rains, the earthworms come to the surface and may be seen on sidewalks. You can collect these for study by turning over the earth with a shovel.

The life-cycles of parasitic worms often involve several different hosts. The tapeworm, for example, may be transferred from the human to a beef and back to the human. The tapeworm is structurally adapted for this life-cycle. Repiratory, circulatory, and digestive systems are lacking in these parasites. They depend on the host for these important processes. The free-living forms, on the other hand, are structurally and behaviorally adapted to a very different environment.

Procedure

Part I. Analysis

Turn to the Data Sheet and look at the diagrams. Use your textbook to identify the planarian, tapeworm, *Ascaris*, earthworm, and leech on the Data Sheet. Answer questions 1–6.

Part II. Investigation

(A) The planarian

Remove a planarian from the collecting jar with a pipette. Then place the planarian in a Syracuse watch glass with some pond water. Examine the animal with a hand lens. Make a current in the water by using a pipette and observe the animal's reaction. The projections on each side of the head are called *auricles*.

Using a penlight flashlight, test the reactions of the animal to light stimulus on various areas of its body. Record your observations in questions 7–10 on the Data Sheet.

Test the response of the animal to food by putting a small piece of liver in the water. After observing the behavior of the planarian, remove the liver. Now put one drop of acetic acid in the water near the planarian. Observe its reaction to this chemical stimulus. If an animal moves towards a stimulus, its reaction is *positive*. If it moves away from a stimulus, its reaction is *negative*. Record your observations in question 11 on the Data Sheet.

Transfer the animal to a glass slide and examine the ventral surface with your hand lens. Describe your observations in question 12 on the Data Sheet. Locate the mouth and the pharynx. Label these on the diagram at the top of the Data Sheet. Label the *auricles*.

(B) The earthworm

Place a live earthworm on a damp paper towel and observe its movements. Can you determine the anterior and posterior end from the way it moves? Compare your specimen to the diagram on the top of the Data Sheet and record your observations in questions 13–14.

Hold the animal in one hand and lightly rub your fingers along its dorsal, lateral, and ventral surfaces. Now examine the animal with a hand lens or dissecting microscope. Determine an explanation for your sensation when lightly rubbing the animal. Record your observations in questions 15–17 on the Data Sheet.

Using the hand lens or dissecting microscope, observe the *dorsal blood vessel* extending along the back of the animal. When you have completed your observations, answer questions 18–19 on the Data Sheet. If you can answer all these questions, your observations were well done. If not, return to the specimen to complete your observations.

Locate the *mouth*. Above the mouth is an overhanging "lip" called the *prostomium*. The hand lens or dissecting microscope will help you to locate the *excretory pores*, which function in the excretion of nitrogen-containing wastes.

Examine the thickened area encircling the body. This is the *clitellum*. It forms a cocoon for depositing eggs. Now answer question 20 on the Data Sheet before continuing your observations.

Test the behavior of the earthworm by carefully observing its reaction to various stimuli such as white light. Put the animal on top of loose, moist soil and observe its behavior. If your observations have been made carefully, you can now record them in questions 21–24 on the Data Sheet. You should also be able to see the following on the earthworm: *mouth, prostomium, anus, setae, clitellum*, and *excretory pores*.

NAME DATE

(C) Internal structures of the earthworm

Using a preserved earthworm, follow these instructions in your dissection of the animal:

1. Read the directions carefully *before* you begin to cut.
2. Identify structures to be cut *before* you begin to cut.
3. Lift the structure to be cut.
4. Cut *only* when directed to do so.
5. When possible, use your fingers *instead* of a needle to expose structures.
6. Before you begin, be sure you understand the meaning of the following terms: *anterior, posterior, dorsal, ventral, median, longitudinal*.
7. In the space provided on the Data Sheet, label the structures. Labeling should be done as you investigate *each* structure.

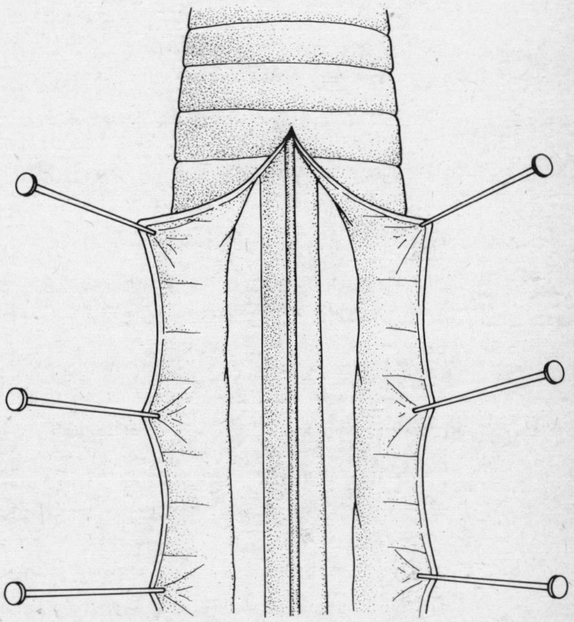

Since most organs lie in the anterior region of the worm, the incision is started 2 cm behind the clitellum and continued forward to the anterior end. The incision should be made to the left of the mid-dorsal line to avoid cutting the dorsal blood vessel. As the body wall is opened, it should be pinned back on both sides as shown. Place the pins about 1 cm apart and at an angle so that they do not obstruct the working area.

Put a preserved earthworm in the dissecting pan, dorsal side up. Pin both ends of the worm to the wax at the bottom of the pan. Use the point of the scissors to make the incision. Use care to cut through the body wall *only*. Extend this incision to the prostomium. Separate the edges of the cut and notice that there is a space between the body wall and the intestine. This is the *coelom*. Observe the partitions (*septa*) extending from the body wall to the intestine. Using forceps and dissecting needles, break the septa. Pin the sides of the body wall to the wax.

The muscular system allows movement of the earthworm. Locate the *circular muscles* just beneath the body wall. Next observe the *longitudinal muscles*, which form a layer just beneath the circular muscles. Think about how these muscles would cause the animal to change shape when they contract. Record your observations in question 25 on the Data Sheet.

The digestive system is composed of a tube extending from the mouth to the anus. The *buccal cavity* passes from the mouth to about the third segment. Observe the thick-walled *pharynx*, which is located in segments 4 and 5.

Follow the long, slender *esophagus* from the pharynx to segment 15. Can you find the large, thin-walled *crop* posterior to the esophagus? Posterior to the crop is another thick-walled organ — the *gizzard*. The *intestine* extends from the gizzard to the *anus*. Referring to your text, answer questions 26–27 on the Data Sheet.

The earthworm is hermaphroditic (*monoecious*). This means that each animal has both male and female reproductive structures. Your text explains how two worms copulate, thus each worm fertilizing the other. The reproductive systems are found in segments 9 to 15. Locate the two pairs of *testes* and the three pairs of white *seminal vesicles* in segments 10, 11, and 12. In segments 9 and 10 are a pair of small, white spherical structures called *seminal receptacles*. Locate the *ovary* beneath the seminal vesicles. Using your text to help you, answer questions 28–30 on the Data Sheet.

The circulatory system can be observed by first locating the *dorsal blood vessel*. It is muscular and forces blood anteriorly. Carefully remove the white seminal vesicles from the left side of the body and separate the septa from segments 6 to 11. Identify the thick, dark *aortic arches* branching from the dorsal blood vessel and passing around the esophagus. These connect with the *ventral blood vessel* below the esophagus. The muscular walls of the aortic arches contract and function as "hearts." Answer question 31 on the Data Sheet.

The excretory system is made up of small, white, loose tubes along each side of the digestive tract. These excretory organs are called *nephridia*. They are located in each segment except the first three and the last one. To see these structures, you may extend the original cut posteriorly about five centimeters. Pin back the body wall, and carefully pull out the intestine. A dissecting microscope and dissecting needle will help you locate the nephridia. Using your text as a reference, answer question 32 on the Data Sheet.

The nervous system can be observed by first examining the area dorsal to the buccal cavity in segments 2 and 3. Find the small white "brain" or *suprapharyngeal ganglion*. Two nerves arise from this ganglion and pass around either side of the esophagus. These nerves join the *subesophageal ganglion* located below the esophagus. From this ganglion, the *ventral nerve cord* extends the length of the animal. Remove part of the intestine and observe the ventral nerve cord. Notice the small ganglion in each segment. Cut the nerve cord and pull it free from the body. Notice the small nerves branching from it. Answer question 33 on the Data Sheet.

Did you observe that a distinct respiratory system was missing? Refer to your text to help you answer question 34 on the Data Sheet. Label the diagram of the earthworm dissection on the previous page.

Investigations On Your Own

(1) You already know that the sponge and hydra can develop an entire animal from a small section. Can planaria do the same? Design and conduct an experiment to test your hypothesis.
(2) Using reference books, compare the life history of a tapeworm (a parasitic worm) to the life histories of the planarian and the earthworm (two free-living worms).
(3) Use your library to find how earthworms are raised. Make a narrow terrarium and fill it with loose earth. Introduce several earthworms. Make observations over a period of time and write a report on your observations.
(4) If your school has specimens of tapeworms available, examine one carefully to identify the structures labeled in your text. Compare the tapeworm to the planarian as to the manner in which it obtains its food, digests its food, absorbs its food, respires, and moves. Research and write a paper on the life cycle of the beef tapeworm, pork tapeworm, fish tapeworm, or dog tapeworm.

NAME DATE

Adaptations of Worms 29

DATA SHEET

Analysis

Label the organism and identify it as a member of the phylum *Platyhelminthes, Nematoda,* or *Annelida*.

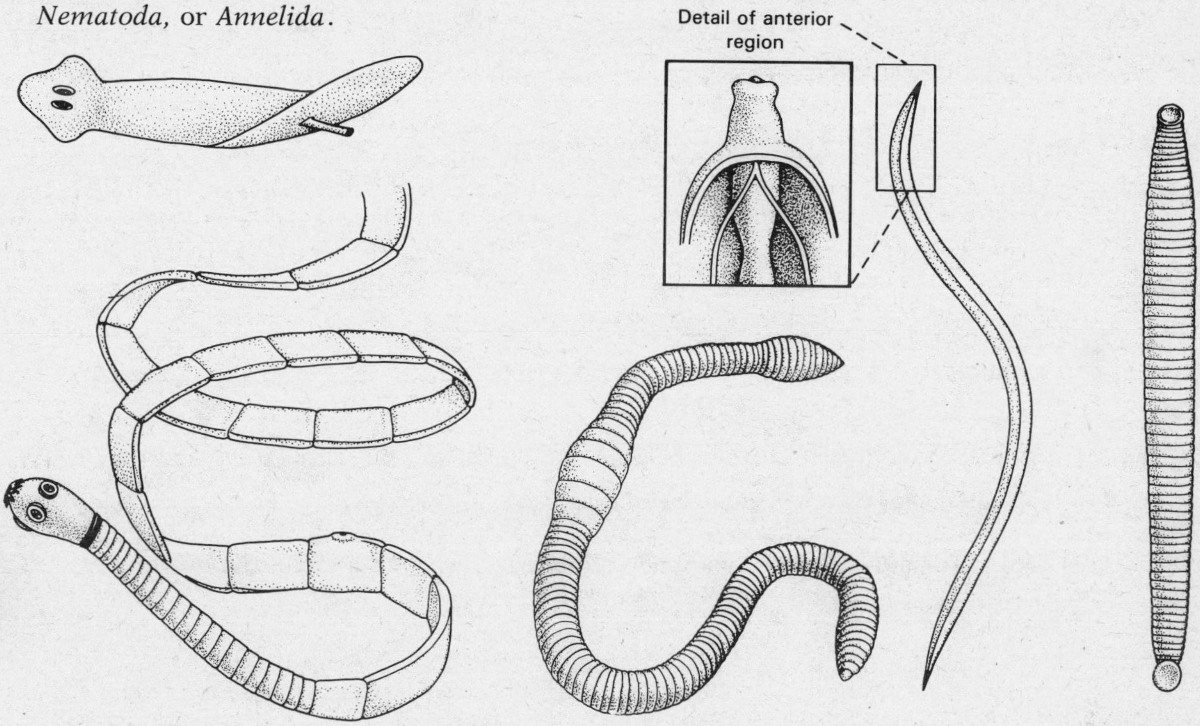

(1) What features indicate the annelids? ______

(2) What features indicate the nematodes? ______

(3) What features indicate the Platyhelminthes? ______

(4) Which of the above worms are parasitic? ______

______

(5) What structure(s) indicate that a worm is parasitic? ______

______

(6) Why are these structures considered to be adaptations to the environment? ______

______

Investigation

(A) The planarian

(7) Briefly describe the dorsal surface of the planarian. ______

(8) Describe the way in which the planarian moves. ______

______

(9) How does the animal respond when light is shined on the eyespots? ______

______

(10) Which end appears to be most sensitive to the light stimulus? ______

(11) Describe the animal's response to the liver. ______

______

(12) Describe the ventral surface of the planarian. ______

______

(B) The earthworm

(13) Describe the earthworm's movements. ______

______

(14) Is there evidence of segmentation? ______

(15) Describe your observations when rubbing your fingers along the earthworm. ______

______

______

(16) How many bristles are on each segment? ______

(17) How do you believe these bristles function in locomotion? ______

______

(18) Does the dorsal blood vessel appear to pulsate? ______

(19) What is the direction of blood flow in this vessel? ______

(20) What is the function of the clitellum? ______

______

(21) Does the earthworm have any visible sense organs? ______

(22) Describe the reaction of the earthworm to light. ______

______

(23) Explain how this behavior is related to survival. ______

______

______

NAME DATE

(24) Record the behavior of the earthworm when placed on top of loose, moist soil.

____

(C) Internal structures of an earthworm: Study the structures carefully and label the following: *pharynx, esophagus, crop, gizzard, dorsal blood vessel, seminal vesicles, intestine, body cavity (coelom), nephridium, ganglion, hearts.*

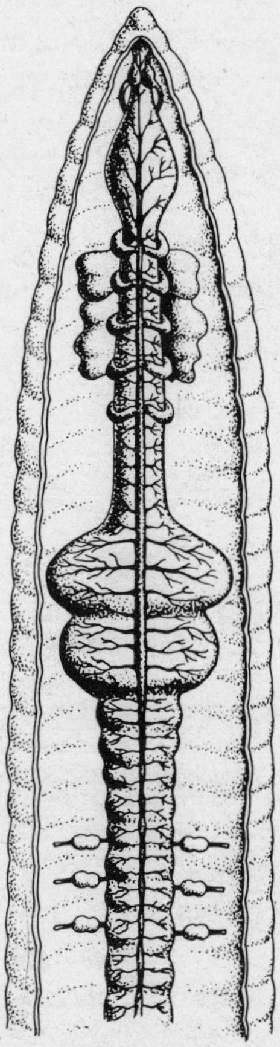

(25) Name the two muscle layers of the earthworm and explain how each functions.

____

____

____

(26) What is the function of the crop? ______________

(27) Explain the function of the thick-walled gizzard. ______________

____

(28) What is the function of the testes in the earthworm? ______________

(29) Where are the sperms stored until they are discharged during copulation? ______

____

(30) What word explains that an organism, such as the earthworm, has both male and female organs in the same animal? ______________

(31) Why is it said that the earthworm has a "closed" circulatory system? ____________

____

(32) Where do wastes leave the body? ____________________________________

____

(33) Where do the small nerves go when they branch from a ganglion of the ventral nerve cord? ____

____

(34) How does respiration occur in the earthworm? ________________________

____

Summary

(1) Do tapeworms, planarians, and earthworms have visible sense organs? ________

____

____

(2) Compare the movement of the planaria and earthworm. ____________________

____

____

____

____

(3) Compare the "feeding" methods of the tapeworm, planarian, and earthworm.

____

____

____

____

(4) Compare the structural adaptations of the parasitic tapeworm to those observed in the free-living planarian and earthworm. ______________________________

____

____

____

____

NAME DATE

30

Phylum Mollusca

MATERIALS

INVESTIGATION▶
- specimens of chiton, clam, snail, and squid
- charts of mollusks
- hand lens
- clam shells
- fresh clam
- dissecting tray
- probe
- screwdriver
- scalpel

OBJECTIVES▶
- To observe the characteristics of the phylum Mollusca
- To observe features of the mollusks that separate them into classes
- To identify and label parts of a clam

Problem

What is a mollusk?
Mollusks are found in marine, fresh-water, and terrestrial environments. All mollusks have a head, foot, and visceral hump. The shell is an important characteristic of mollusks and is used to separate the phylum into its classes: *Pelecypoda, Gastropoda,* and *Cephalopoda.*

Procedure

Part I. Analysis

Examine the four diagrams of the three classes of Mollusca on the Data Sheet. Identify and compare the head, foot, and visceral hump of each to the others. Write in the *class* of each. Identify and label: *anterior, posterior, ventral*, and *dorsal*.

These are characteristics common to all the adult organisms. Closely related organisms also develop in similar ways. The "blueprint" for development is passed on to the zygote by the DNA of the gametes. All mollusks pass through a larval stage called a *trochophore*. As the trochophore grows, it begins to develop the characteristics that identify its specific kind. Using your text to help you, answer questions 1–4 on the Data Sheet.

Part II. Investigation

A. Comparison of various mollusks.
Examine specimens or use charts or diagrams in this Investigation. Identify the type of symmetry. Examine and compare those specimens with external shells, and observe the thickness and arrangement of the shells. Examine the muscular foot of each mollusk. Then compare the heads. Notice that the head is most highly developed in the squid. The large eyes provide a clue about the activities of these mollusks. If you have a specimen of a chiton, you will have to examine its ventral surface to find its very small head, indicated by the location of the mouth. Now, compare the snail's head to that of the chiton. Answer questions 5–12 on the Data Sheet to record your observations.

B. Observing a clam.
Place a clam in a dissecting tray and examine it. Find the *elastic ligament* on the *dorsal* surface. This fastens the two shells (or valves) together. Determine the anterior and posterior ends of the animal. The hump on the shell is anterior to the ligament. This is the *umbo* and is the oldest part of the shell.

As the clam grows, it secretes layers; each extends beyond the last. This makes a series of concentric *growth rings*. Find the growth rings on the clam shell. Answer questions 13 and 14 on the Data Sheet.

Now, examine an empty clam shell. Notice that it is made up of three layers. The thin, outside layer is the *horny layer*. Look at a broken piece of the clam shell with your hand lens. Observe the middle or *prismatic layer*. The inner layer of a clam shell is the *pearly layer*. Record your observations in questions 15–17 on the Data Sheet.

Carefully open the shells of your clam specimen with a screwdriver. Use a scalpel to cut the two large muscles that hold the shell closed. Open the shell and lay it back on the dissecting tray. Examine and identify the muscles: *anterior* and *posterior adductor muscles*. Observe the *scars* that are left on the shell by the removal of these muscles. Notice the toothlike projections on the dorsal edge of the shell. These fit into grooves in the opposite shell. Answer question 18 on the Data Sheet.

Notice the mantle that surrounds the animal and lines the shell. The space between the two parts of the mantle is the *mantle cavity*. You can examine the organs in the mantle cavity by removing part of the mantle with a scalpel. Observe the sheetlike *gills* in the mantle cavity. Notice the *muscular foot*. Record your observations in question 19 on the Data Sheet.

To observe how water circulates through the mantle cavity, notice two joined tubes (*siphons*) on the posterior end of the animal. The *incurrent siphon* is the ventral one and brings water into the mantle cavity. The *excurrent siphon* is the dorsal one and removes water from the mantle cavity. When water enters the mantle cavity, it flows over the gills, which function in respiration. Explain your conclusions in question 20 of the Data Sheet.

The gills are adapted to perform another important function. Try to find *cilia* on the gills with the aid of a hand lens or a dissecting microscope. When water passes over the gills, mucus traps particles and the cilia move them. A pair of *palps* on either side of the stomach directs the particles toward the mouth. Locate the *mouth* between the anterior end of the palps.

Using the scalpel, cut away the muscle of the foot to locate the *digestive gland* that surrounds the posterior part of the stomach. Identify the *intestine* and follow it to where it passes through the *pericardial cavity*. You will then be able to identify the *heart*. Fill in the chart, question 21.

Investigations On Your Own

(1) Fill a narrow aquarium with sand and then add water to a depth of about six centimeters above the sand. (Use salt water or fresh water depending on the habitat of the animal you will use.) Place freshly collected clams on the surface and observe the method by which they dig. Once they appear to be siphoning water, add a little lampblack or carmine to the water and observe the action of the currents. After they have siphoned water for a few hours, dissect an animal and locate the carbon or carmine particles inside the animal. Write up your observations.

(2) Measure the speed of a snail's pace. On a piece of paper, draw concentric circles at 1-centimeter intervals. Place the white paper under a glass container and put the snail in the center. When it starts to move, measure the distance it travels in 30 seconds. Repeated tests will allow you to obtain an average speed for "snail travel."

Test the speed under various conditions, such as temperature, surface, and slope. Record your data in charts and write up your observations.

NAME ____________________ DATE ____________

30

Phylum Mollusca

DATA SHEET

Analysis

Examine these four diagrams of the three classes of Mollusca. Identify and compare the head, foot, and visceral hump of each to the other. Write in the class of each. Label: *anterior, posterior, ventral,* and *dorsal*.

Clam	*Snail*	*Chiton*	*Squid*
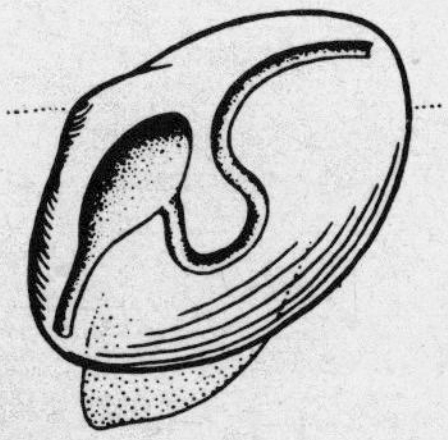	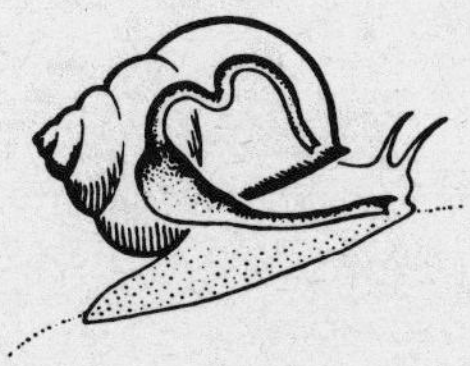		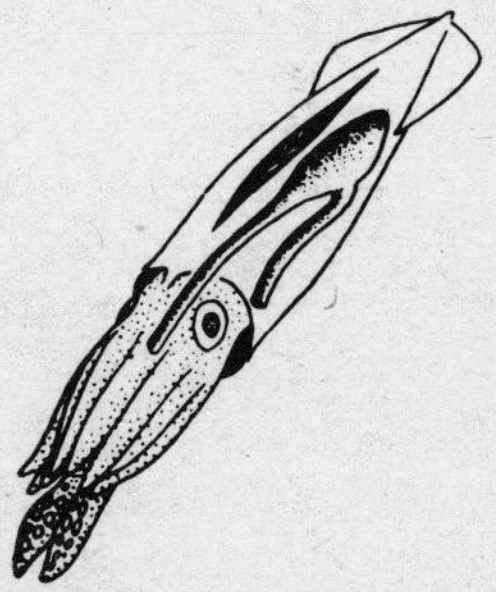

(1) What similar characteristics can you observe in the adult mollusks? ____________

____

____

(2) At what developmental stage would each of the above mollusks look alike? ______

____

(3) Describe the larval form of the mollusks. ______________________

____

____

____

____

____

(4) If you were examining water under a microscope and saw one of these interesting larvae, could you be certain it was a mollusk? Explain your answer.

____

____

Investigation

A. Comparison of various mollusks

(5) Which ones have an external shell? ______________________________

(6) What symmetry does each have? ______________________________

(7) Do you observe any evidence of segmentation on the specimens? ______________

____

____

(8) Compare the foot of the clam, snail, and chiton. ______________________________

____

____

(9) What structures form the foot of the squid? ______________________________

____

(10) Are jointed appendages present on any of your specimens? ______________

(11) From your observations, what characteristics do all the mollusks possess? ________

____

(12) What features are different enough to separate the mollusks you have observed into three classes? ______________________________

____

B. Observing a clam

(13) How many shells (or valves) are present in the clam? ______________

(14) Can you now determine the right and the left shell of a clam? ______________

(15) Describe the appearance of the horny layer of a clam shell. ______________

____

(16) Describe the appearance of the prismatic layer of a clam shell. ______________

____

(17) Describe the appearance of the pearly layer of a clam shell. ______________

____

(18) What purpose is served by the toothlike projections on the dorsal edge of a clam shell? ______________________

____

(19) Where is the muscular foot located? ______________________

____

(20) What gas exchanges occur in the gills? ______________________

____

(21) Draw a diagram of the internal structures of the clam and label the following on the diagram below: *hinge, umbo, mouth, esophagus, stomach, intestine, anus, digestive gland, heart, gills, foot, anterior adductor muscle, posterior adductor muscle, incurrent siphon, excurrent siphon*.

Summary

(1) Compare the heads of the clam, chiton, snail, and squid. ______________________

____

____

____

____

____

____

(2) What structures of the mollusks distinguish them, as a group, from other phyla you have studied? ______________________________

____

____

(3) What structures are used to distinguish the classes Pelecypoda, Gastropoda, and Cephalopoda? ______________________________

____

(4) Explain how a bivalve, like the clam, obtains its food. ______________________________

____

____

____

____

(5) What is the habitat of various mollusks? ______________________________

____

____

NAME DATE

31 Crayfish Adaptations

MATERIALS

INVESTIGATION▶

- live crayfish
- small rock
- probe
- pencils (red, green, yellow, black)
- penlight flashlight
- India ink
- diagrams or charts of crayfish anatomy

OBJECTIVES▶

- To examine the external adaptations of a crayfish
- To observe behavior of a crayfish

Problem

What structural and behavioral adaptations can be observed in a crayfish?

Crayfish live in nearly all bodies of fresh water that contain lime. They dig holes in banks, and the water in which they live is often murky.

The segmented appendages of the crayfish carry on several different functions. The swimmeret is considered the basic type of crustacean appendage. Appendages that are similar in origin and structure, but have a different function, are referred to as homologous structures.

Procedure

Part I. Analysis

Turn to the Data Sheet. Using your colored pencils and your text to help you, identify the parts of the digestive, circulatory, nervous, and reproductive systems.

Part II. Investigation

Put a living crayfish in the aquarium or a large fingerbowl and cover it with water. Notice that the body of the crayfish is divided into two distinct regions: the *cephalothorax* and the *abdomen*. Observe the *carapace* covering the cephalothorax. The abdomen ends in a reduced abdominal segment called the *telson*.

Notice the *cervical groove* that marks the division between the head and the thorax. Observe that the carapace extends forward. This forms a horny beak called the *rostrum*. Notice the location of the eyes. Record your observations in questions 1–3 on the Data Sheet.

The most anterior appendages are the branched *antennules*. Locate the *antennae*, which are attached to the head, posterior to the antennules. Record your observations in question 4 on the Data Sheet.

Now, arrange the small aquarium so you can observe the mouthparts of the crayfish. Record your observations in question 5 on the Data Sheet.

The thorax has eight pairs of appendages. Three pairs are *maxillipeds* and are used in handling food. Five pairs are legs. Try to identify all eight pairs. Compare the maxillipeds to the walking legs. Notice that the first pair of legs are well-developed *chelipeds*. The next four pairs of legs are called *walking legs*. Record your observations in question 6 on the Data Sheet.

Arrange the aquarium so you can observe the ventral surface of the abdomen.

The abdominal appendages of the crayfish are called *swimmerets*. In the female, the first pair of swimmerets are small. In the male, the first two pairs are modified for transferring sperms. Determine the sex of your specimen.

The sixth pair of swimmerets are enlarged to form the *uropods*. The uropods and the telson form the powerful tail fin used in backward swimming. Find the *anal opening* on the ventral side of the telson.

Now, place a rock in the bottom of the aquarium. Observe the behavior of the crayfish for 5 minutes and record your observations in questions 7 and 8 on the Data Sheet.

Touch the antenna with a probe and observe its reaction. Answer question 9.

Shine the penlight flashlight on the anterior end of the crayfish. Then shine it on the posterior end. Observe the reactions. Answer question 10.

Put 5 drops of saturated salt solution in the water near the animal and observe the results. Answer question 11.

Clean the aquarium and add a drop of India ink just posterior to the cephalothorax. Carefully observe the movement of the ink particles. Explain your observations in question 12 on the Data Sheet. Then, on the diagram of the crayfish at the beginning of the Data Sheet, label the following: *rostrum, antennule, antenna, compound eye, cephalothorax, abdomen, cheliped, telson, uropod*, and *walking leg*.

Investigations On Your Own

(1) You may study adaptive behavior of sowbugs or pillbugs. These are common terrestrial crustaceans. Watch them walk and test their responses to various stimuli.

Put 5 or 6 animals in two dishes. Place a piece of wet paper towel in one. Keep the dishes away from heat or strong light, and observe them for 10 or 15 minutes. Compare the results.

Cut a piece of paper towel to fit the bottom of 2 dishes. Moisten the paper and put 5 or more animals in each dish. Cover one-half of each container with black lightproof paper. Put one container in a closed drawer or under a box. Put the other container in a well-lighted area. Compare the results after 10 or 15 minutes.

Line the bottom of 2 dishes with black, lightproof paper. Cover the paper with a damp paper towel and place 5 or more animals in each. Now, place a light under one side of one dish. Compare the results after 10 or 15 minutes. Summarize your conclusions about the reactions of sowbugs and pillbugs to moisture, light, and heat. When do you think these animals would search for food? Do your experiments suggest the habitat of these animals?

(2) Make diagrams of sowbugs and pillbugs. Compare the characteristics of these terrestrial crustaceans to the crayfish. What structural adaptations allow the sowbugs and pillbugs to live on the land?

NAME DATE

Two Common Insects 32

MATERIALS

INVESTIGATION▶

- living flies
- Petri dishes
- honey or sugar solution
- preserved grasshoppers
- charts
- dissecting pan
- scissors
- forceps
- scalpel
- hand lens

OBJECTIVES▶

- To identify characteristics of the class Insecta
- To identify characteristics of two orders
- To identify specialized adaptations in the grasshopper
- To examine the complex feeding structures of the grasshopper
- To observe the organ systems of a grasshopper

Problem

In what ways are flies and grasshoppers alike, yet unalike?

Flies and grasshoppers are common examples of the class Insecta. Flies, mosquitoes, and gnats belong to the order Diptera. Grasshoppers, crickets, katydids, locusts, and cockroaches belong to the order Orthoptera.

Insects breathe through tiny tubes called *tracheae*. The number of body regions, structure of mouthparts, pairs of antennae, structure of legs and wings are all used to identify insects.

Grasshoppers and houseflies share many characteristics. These similarities are recognized in the class Insecta. Their differences place them in separate orders. The life history of each order of insect is different. A grasshopper's egg may remain over winter and hatch into a nymph in the spring. The nymphs resemble the adults but do not have functional wings or reproductive systems. Since the nymphs resemble the adults, the process is called *incomplete metamorphosis*. Usually one generation is produced each year.

The housefly's egg may hatch into a larva within a few days. The larval form, called a maggot, will then form a pupa and change into an adult within two weeks. The process is *complete metamorphosis*. Several generations of houseflies develop every year.

Procedure

Part I. Analysis

Turn to the Data Sheet and examine the diagrams of the housefly and the grasshopper. Compare and identify the stages of the life histories of these animals. Using your text to help you, identify these stages and the characteristics of the class Insecta. Answer questions 1–6 on the Data Sheet.

Part II. Investigation

(A) Comparison of the adult housefly and the grasshopper

Place a drop of honey or sugar solution in the bottom of a Petri dish and introduce a fly. Carefully observe the fly's behavior. Lift the dish, hold the hand lens to the bottom, and observe the lapping mouthparts of the fly. Identify the *3 body regions, 1 pair of antennae,*

3 pairs of legs, and *2 pairs of wings*. Locate the large *compound eyes*. Label the *lapping mouthparts* and *compound eyes* on the diagram on the Data Sheet.

If living specimens are available, observe the way in which grasshoppers move. Compare the legs and wings to those of the fly.

Place a preserved grasshopper in a dissecting pan, and identify the characteristics of the class Insecta. Compare your observations to those of the fly. Can you determine the function of the mouthparts? Record your observation in 7–9 on the Data Sheet.

Use your hand lens to examine your specimen more carefully. Your text and charts may also be helpful. Add the following labels to the diagram on the Data Sheet: (structures on the head) *compound eye, simple eye, mouthparts;* (structures on the thorax) *walking legs, leaping legs, forewing, hindwing;* (structures on the abdomen) *spiracles, ovipositor*.

Examine the mouthparts of the grasshopper to see how they are adapted to feeding on grass. Use the hand lens. Carefully remove each mouthpart with a forceps. Compare the mouthparts to those in the figure below. Record your conclusions about your observations in 10–14 on the Data Sheet.

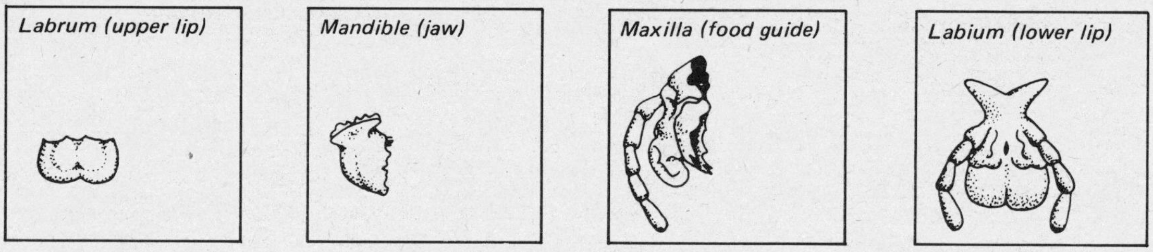

(B) Internal anatomy of the grasshopper

Read the directions carefully *before* beginning your dissection. The organs you are to find are indicated in **boldface**.

First, remove the three left legs. Insert the point of your scissors under the dorsal surface of the last abdominal segment. Being careful not to cut underlying organs, make an incision just to the left of the mid-dorsal line. Continue the cut through the thorax. Then, cut down the left side to the ventral surface in front of the thorax. Make another lateral cut on the left side between the next-to-the-last and the last abdominal segments.

With the forceps, pull down the left side and look for the large **dorsal blood vessel** ("heart"). Now, use your scalpel to cut the muscles close to the exoskeleton as you slowly pull down the left side of the insect. Locate the finely branched **tracheae** going to the spiracles.

Insert the point of the scissors at the back of the head and cut through the exoskeleton over the top of the head between the left antenna and left eye. Extend this incision to the **mouth**. Use the forceps and scalpel to remove the exoskeleton on the left side of the head. Locate the **dorsal ganglion** ("brain"). You may be able to identify a large, whitish nerve passing from the dorsal ganglion around the **esophagus** to the **ventral ganglion**. Follow the ventral nerve cord posteriorly and compare your observations to diagrams in your text or available charts. You should locate several **ganglia** in your specimen.

Cut away the tissue to expose the digestive tract along the length of the animal. From the posterior end of the esophagus, the digestive tract enlarges into a **crop**. Posterior to the crop, the wall of the digestive tract becomes muscular and is the **gizzard**. Notice that the gizzard and the **stomach** are separated by a narrowing of the digestive tract. Also observe that many **gastric caeca** are attached here.

Posteriorly, another narrowing separates the stomach from the **intestine**. Identify the many long, thin, **Malpighian tubules** attached at this point. Note the narrow **colon** posterior to the intestine. The digestive tube enlarges to form the **rectum**. Here, wastes are concentrated before passing out through the **anus**.

Note the **ovary** in the female or the **testes** in the male. Label the structures you have observed on the Data Sheet, and answer questions 15 and 16.

NAME DATE

Two Common Insects 32

DATA SHEET

Analysis

Using your text to help you, identify the development of each life history by labeling: *complete metamorphosis* and *incomplete metamorphosis*. On the appropriate diagram, label: *egg, nymph, larva, pupa,* and *adult*. On the separate diagrams of the adults, label: *anterior, posterior, dorsal, ventral, head, thorax, abdomen, antennae, legs*, and *wings*.

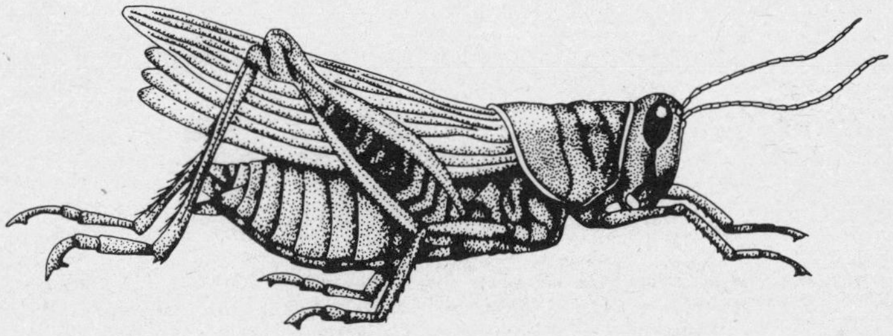

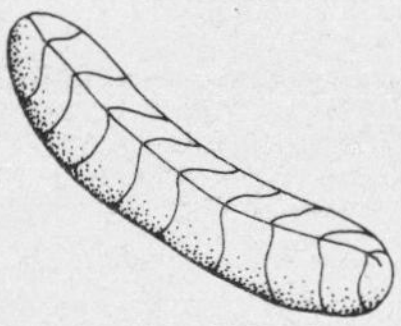

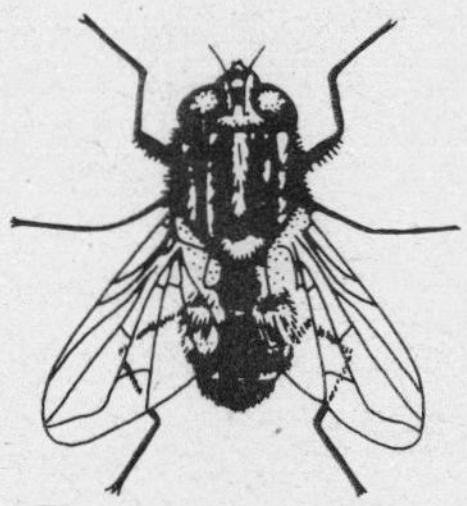

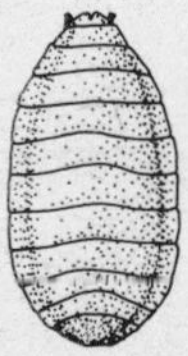

(1) What are the stages of incomplete metamorphosis of the grasshopper? ____________

____

(2) How does the nymph differ from the adult grasshopper? ____________

____

(3) Do the nymphs and adults eat the same food? ____________

(4) What are the stages of complete metamorphosis of the housefly? ____________

____

(5) How does the larva differ from the adult housefly? ____________

____

____

____

____

(6) Do the larvae and adult houseflies eat the same food? ____________

Investigation

(A) Comparison of the adult housefly and the grasshopper

(7) What characteristics are shared by the housefly and the grasshopper? ____________

____

____

(8) State the order of, and the characteristics observed in, the housefly. ____________

____

____

(9) State the order of, and the characteristics observed in, the grasshopper. ____________

____

____

____

(10) What is the function of the labrum? ____________

____

(11) How are the mandibles adapted for crushing blades of grass? ____________

____

(12) How are the maxillae adapted for holding and cutting food?

(13) What is the function of the labium?

(14) How do the palpi of the maxillae and the labium function in feeding?

(B) Internal anatomy of the grasshopper

Compare the diagram of the dissected grasshopper below with your observations and label: *labrum, labium, esophagus, crop, gizzard, gastric caeca, stomach, Malpighian tubules, intestine, colon, rectum, anus, "heart," dorsal ganglion, ventral nerve cord*. In the diagram, draw the ovaries or testes you observed.

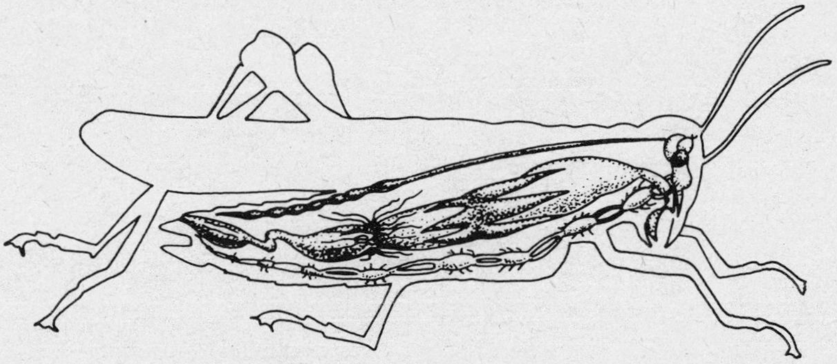

(15) What is the function of the gastric caeca?

(16) What is the function of the Malpighian tubules?

Summary

(1) What might be an advantage and a disadvantage of incomplete metamorphosis to an insect such as the grasshopper?

(2) What might be an advantage of complete metamorphosis to an insect such as the fly?

(3) How do the structural differences in the grasshopper and housefly enable them to exhibit more complex behavior than centipedes and millipedes?

(4) After comparing the structural differences between the insects (such as houseflies and grasshoppers) and the spider, what conclusions might you state regarding the range of distribution of an individual animal and of the species?

NAME DATE

33 Vertebrate Adaptations

MATERIALS

INVESTIGATION▶

- Living or preserved specimens of:
  - lamprey eel
  - dogfish shark
  - fish
  - frog
  - turtle
  - snake
  - bird
  - mouse or gerbil
  - (any other available specimens may be used or substituted)

OBJECTIVES▶

- To determine how body shape can be adaptive
- To observe differences in vertebrate appendages
- To compare vertebrate head structures and sense organs
- To observe differences in body coverings

Problem

What characteristics can be observed in a variety of vertebrates?

The key word in the above question is *vertebrates*. A vertebrate is an animal with a *backbone* made up of bone and/or cartilage. An anterior *pectoral girdle* and a posterior *pelvic girdle* are suspended from the vertebral column. Various types of limbs or appendages are attached to the girdles. The variety of structures are adaptations to certain environments.

Procedure

Part I. Analysis

Before examining structural differences among the various vertebrates, identify several common characteristics of *all* vertebrates. Using your text to help you, turn to the Data Sheet and label the parts of the cat skeleton. Then answer questions 1–3.

Part II. Investigation

Assemble the specimens in the order listed above. Observe that water, land, air, and combinations of these environments are occupied by these vertebrates. Using your text to help you, answer questions 4 and 5 on the Data Sheet.

Now observe and compare the heads and bodies of your specimens. Answer question 6 on the Data Sheet and record the habitat for each representative in the chart on the last page of the Data Sheet.

Observe the general shape of the specimens. Notice that some are streamlined. Answer question 7–8 on the Data Sheet, and then record the body shape of each animal in the chart on the Data Sheet.

Observe and compare the appendages of the specimens. *An appendage is any part of the body attached to the main body trunk, i.e. arms and legs*. Appendages are used primarily for locomotion, although certain minor functions are also performed by them. Consider the environment of each animal and determine the function of the appendages. Notice that the appendages of some animals have digits. A digit is a finger or toe. Then, note the appendages and digit structure in the chart. Indicate whether the animal has fins and appendages, and, if digits are present, whether they are clawed.

Examine the head of each animal and compare the shapes. Look for the presence or absence of: *eyes, eyelids, irises of the eyes, ears* or a structure for hearing, *nostrils, mouth structures, tongue,* and *teeth*. If you are observing living specimens, try to gather evidence for sensory development by using various devices such as food, brightly colored objects, etc. Try to make a judgment on the apparent sensory development of each animal. Answer question 9 on the Data Sheet. Then record your observations under the proper heading in the chart. You may use the following symbols for recording sense organs:

(0)	— not present or unable to be determined
(+)	— present, poorly developed
(++)	— fairly well developed
(+++)	— well developed

Carefully observe the skin and body covering of each specimen. Compare them, and answer questions 10–11 on the Data Sheet. Record your observations on the chart.

Investigations On Your Own

(1) Compare the skeletal features of four vertebrate animals. Observe the structure of the backbone, ribs, hip and shoulder girdles, limbs, skull, and jaws. Note the skeletal adaptations for locomotion, protection of the vital organs, and body shape. Summarize your observations in a table.

(2) Visit a zoo and record the name of each animal you study. Note its original habitat and make a list of unusual characteristics you actually observe. Determine how these characteristics might be adaptive for the animal. Keep in mind that the actual behavior of these captive animals is not the same as it is when they are in their natural habitat.

(3) An interesting study would be to select a captive animal in a zoo and observe its behavior for a period of time. Visit the animal at various times of the day and keep records on its behavior. By reading about this animal, you will be able to make some interesting conclusions about how it has adapted to a completely different environment. You might ask the zookeeper about the nutritional aspects of special diets and the provision of other special requirements to maintain the animal in a healthy condition.

NAME DATE

Vertebrate Adaptations 33

DATA SHEET

Analysis

Vertebrates have an internal framework called an *endoskeleton* made of bone and/or cartilage. In the diagram of the cat skeleton below, label: *anterior, posterior, ventral, dorsal, axial skeleton, pectoral girdle, pelvic girdle, appendicular skeleton, head, trunk, tail.*

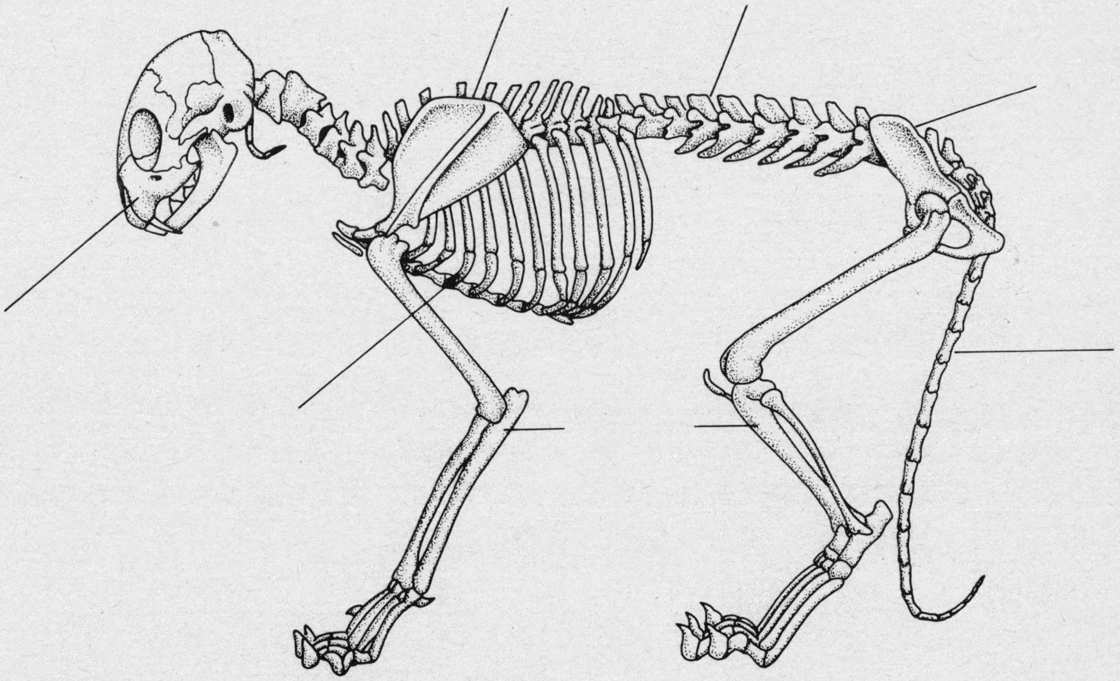

(1) What kind of skeleton does the cat have? ______________________

(2) What functions are performed by this type of skeleton? ______________________

____

____

(3) What is the function of the pectoral and pelvic girdles? ______________________

____

____

Investigation

(4) What adaptations are found in the vertebrate circulatory system? ______________________

____

____

(5) What is characteristic about the vertebrate nervous system? ______

______

______

(6) From observing the specimens, why do you believe they are all vertebrates? ______

______

______

(7) What does it mean for a body to be streamlined? ______

______

______

(8) How does streamlining relate to the habitat of the animals? ______

______

(9) Why is it important to the animal that sense organs be located at the head end?

______

(10) Describe how the scales of various organisms differ. ______

______

______

______

(11) How can the presence of hair be considered an adaptation? ______

______

______

Summary

Using the data obtained from comparing the animals, summarize how each of the following allows for adaptation to a particular environment.

(1) Body Shape ______

______

______

______

______

______

NAME DATE

Animal and Class	Habitat	Body Shape	Appendages	Head-Mouth Structures	Sense Organs	Body Covering
Lamprey Eel Cyclostomata					H—hearing S —sight T —taste Sm—smell	
Dogfish shark Chondrichthyes					H S T Sm	
Bony fish Osteichthyes					H S T Sm	
Frog Amphibia					H S T Sm	
Turtle and snake Reptilia					H S T Sm	
Bird Aves					H S T Sm	
Mouse or gerbil Mammalia					H S T Sm	

(2) Appendages and Digits ______

______

______

______

______

(3) Head and Sense Organs ______

______

______

______

______

(4) Body Covering ______

______

______

______

______

In each of the following questions, a characteristic is given. Can you write in the name of the class?

(5) Members of this class possess a rounded mouth. ______

(6) Members of this class possess scales and clawed digits. ______

(7) In what class would you find appendages modified for flight? ______

(8) Which class is characterized by having a body covering of hair? ______

______

(9) Which classes of vertebrates have fins? ______

______

(10) Members of which class support themselves entirely on their hind limbs? ______

______

(11) The most highly developed sense organs could be found in which class? ______

______

(12) Which class is characterized by having a smooth body covering and the ability to live on land? ______

NAME DATE

Fishes, Aquatic Vertebrates 34

MATERIALS

INVESTIGATION▶

- fresh or preserved yellow perch (black sea bass or other carnivorous species may be used)
- microscope
- slide
- cover slip
- dissecting pan
- dissecting microscope
- probe
- forceps
- scissors

OBJECTIVES▶

- To examine the external structures of a fish
- To dissect and observe the internal structures of a fish
- To understand how these structures are adaptive

Problem

What adaptations for an aquatic life can be observed in fish?

Bony fishes belong to the class Osteichthyes. They have the vertebrate characteristics you observed in the last Investigation. Their streamlined shape and fins adapt them for moving through the water. Their body covering of scales allows water to flow smoothly over the body surface.

Fishes have other adaptations to an aquatic environment to aid in catching prey, escaping enemies, and maintaining their position at a particular depth. Other adaptations allow fishes to remove dissolved oxygen from water.

Procedure

Part I. Analysis

Turn to the Data Sheet to find drawings of three different kinds of fishes. Look at these and answer questions 1–5.

Part II. Investigation

(A) External anatomy of a bony fish

Obtain and place a fish in a dissecting pan. Examine it to determine its body regions. Locate the *head, trunk,* and *tail*. Open the mouth and observe the action of the upper jaw (*maxilla*) and the lower jaw (*mandible*). Examine the teeth and tongue. Insert a probe into one nostril to determine where it leads. On the diagram on the Data Sheet, label: *head, trunk, tail, maxilla, mandible,* and *nostril*. Answer questions 6–8.

Open the mouth wide and probe into the throat cavity (or *pharynx*). Raise the *operculum* covering the *gills*. On the Data Sheet, label the *operculum* and answer question 9.

Examine the *esophagus* with the probe. In the living fish the esophagus is elastic. Record your conclusions by answering question 10 on the Data Sheet.

Observe the narrowed extension of the body attaching the head to the trunk. This is called the *isthmus*. Answer question 11 on the Data Sheet.

Examine the dorsal fin by pulling it forward. Notice that it consists of a spiny portion and a soft ray portion. These are the *anterior dorsal fin* and the *posterior dorsal fin*. Label them on the diagram on the Data Sheet. Answer question 12.

Examine all other fins of your specimen and determine their structure, number, location, probable function, and whether they are supported by spines or spines and rays. Record your observations in the table on the Data Sheet. Label the diagram with: *pectoral fin, pelvic fin, anal fin, caudal fin*.

The *lateral line* is believed to be a pressure-sensitive organ. It is a row of pitted scales along the sides of the body. Label it on the diagram on the Data Sheet.

Examine the scales on your specimen. Note that they overlap like shingles on a roof. Using forceps, remove several scales. Rinse a scale in water and make a wet mount, using a cover glass to keep it flat. Examine it with low power of a microscope.

As a fish grows, the size, rather than the number, of scales increases. As the scales grow, concentric rings form. Notice that the rings are unevenly spaced. Count the regions of closely spaced concentric rings formed during the winter and determine the number of winters your specimen has lived. Answer questions 13 and 14 on the Data Sheet.

(B) Internal anatomy of a bony fish

Locate the gills and observe the gill rakers. Answer questions 15–17 on the Data Sheet.

Read the following directions carefully before starting. Hold the fish, ventral side up, with the head pointing away from you. Insert the point of your scissors through the body wall anterior to the anal opening. Cut forward to the space between the gill covers on the lower side of the head (cut 1). Now lay the fish with its head to the left in the dissecting pan. Continue the incision from the isthmus to the top of the body cavity (cut 2). Make another incision close to the anal opening (cut 3). Carefully remove the muscular body wall from underlying organs. This creates a flap of muscular tissue that can be removed by using a scalpel to cut the attached edge along the top edge of the body cavity (cut 4). The organs of the body cavity should now be exposed.

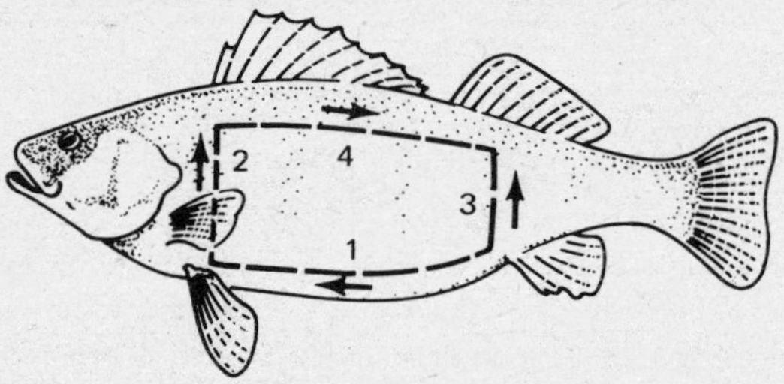

Procedure for opening body cavity

Locate the reddish-brown *liver* (cream-colored in preserved specimens). On the lower side of the liver, locate the *gall bladder*. Remove the liver to expose the short *esophagus* and the *stomach*. Locate the *pylorus* where the stomach and intestine join. Note the small, pouch-like structures (*pyloric caeca*) extending from the pylorus. Follow the loops of the intestine to the *anus*. It may be necessary to remove masses of fat. Cut the alimentary canal at the esophagus and just above the anal opening and remove it.

Locate the *gonads*. The female will have a large, yellowish *ovary* containing many eggs (known as *roe*). In the male, *testes* are found in the same location. The testes are somewhat smaller than the ovary and creamy white. Locate threadlike ducts leading from the gonad and a small tube leading from the *urinary bladder*. Both join and lead to the *urogenital opening* just posterior to the anal opening.

The *air bladder* lies along the top of the body cavity. It may have been broken in removing the body wall. Locate dark masses of tissue (*kidneys*) along the spine.

Section the isthmus carefully to expose the *heart* lying in the *pericardial cavity*. Locate the large *cardinal vein*, which carries blood from the body to the heart. It empties into a thin-walled sac, the *sinus venosus*, which is located at the top of the heart. Find a soft-walled upper chamber, the *atrium*. Below and anterior to the atrium is the thick-walled muscular *ventricle*. A muscular enlargement, the *bulbus arteriosus*, leads from the ventricle to the *ventral aorta* and to the gill arches. On the drawing of the fish heart on the Data Sheet, label: *cardinal vein, atrium, sinus venosus, bulbus arteriosus, ventricle, ventral aorta*.

NAME DATE

34 Fishes, Aquatic Vertebrates

DATA SHEET

Analysis

The drawings below illustrate three different kinds of fishes. The *Cyclostomata* (or *Agnatha*) includes the lamprey. The *Chondrichthyes* includes the sharks, rays, and skates. The *Osteichthyes* includes the bony fishes. Label the drawings with these names, and then label: *anterior, posterior, dorsal, ventral*.

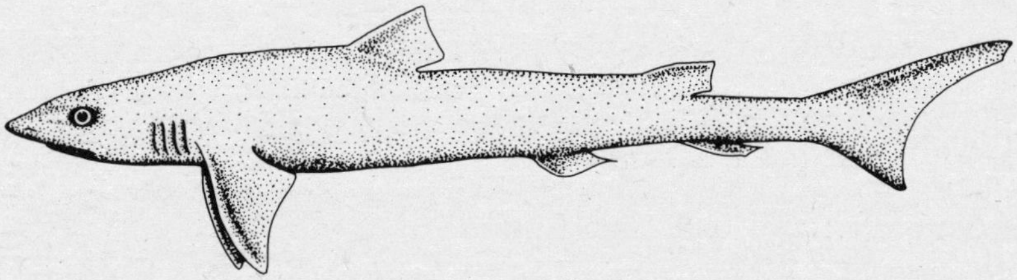

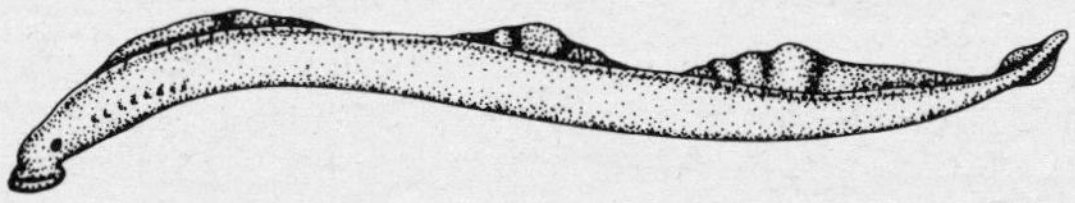

(1) How do the fins of the lamprey differ from those of the other fishes? ______

______

(2) How do the fins of the shark differ from the fins of the bony fish? ______

______

(3) What does the structure of the lamprey's mouth tell you about its eating habits?

______

(4) What does the position of the shark's mouth tell you about its eating habits?

______

(5) What do all three kinds of fish have in common? ______

______

Investigation

(A) External anatomy of a bony fish

(6) Describe the teeth of your specimen and explain how the structure is adaptive to its diet. ______________________________

____

(7) Based on its structure, location, and attachment, determine the probable use of the tongue: ______________________________

____

(8) Describe the location of the nostrils and explain where they lead. ______________

____

____

____

____

(9) Into what structure does the pharynx lead? ______________

(10) Of what adaptive value is the large size and elasticity of the esophagus? ________

____

(11) How is the structure of the isthmus important to the functioning of the gills?

____

____

(12) Suggest a function for the spiny anterior dorsal fin. ______________

Name of Fin	Number	Location	Support	Function
Pectoral				
Pelvic				
Dorsal				
Anal				
Caudal				

(13) Describe the structure of the embedded scale edge and the portion of the scale that overlaps it. ______________________________

____

NAME DATE

(14) What is the approximate age of your fish? ______________________

Label: *head, trunk, tail, maxilla, mandible, nostril, operculum, anterior dorsal fin, posterior dorsal fin, pectoral fin, pelvic fin, anal fin, caudal fin, lateral line.*

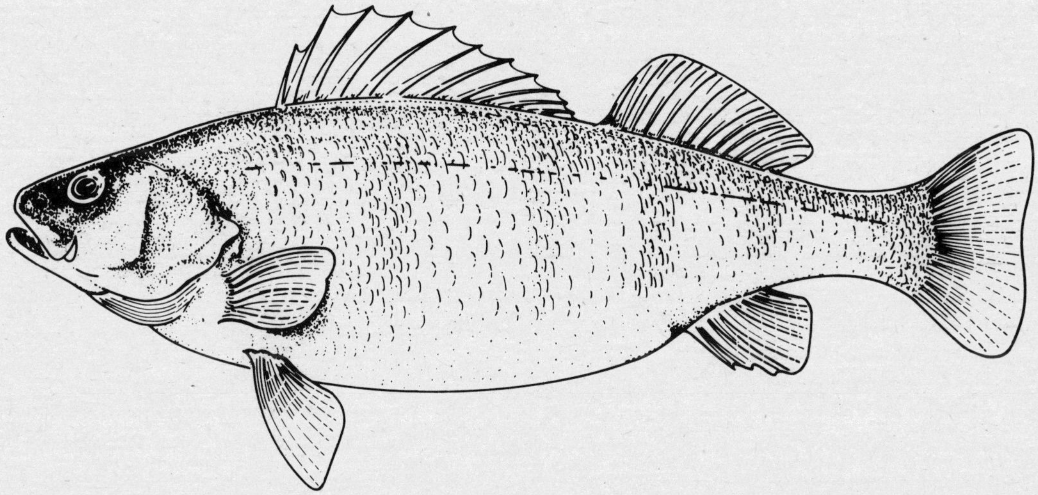

(B) Internal anatomy of a bony fish

(15) How many gills are on each side of your specimen? ______________________

(16) What function takes place in the filament? ______________________

____

(17) What is the function of the gill rakers? ______________________

____

____

Label: *cardinal vein, atrium, sinus venosus, bulbus arteriosus, ventricle, ventral aorta.*

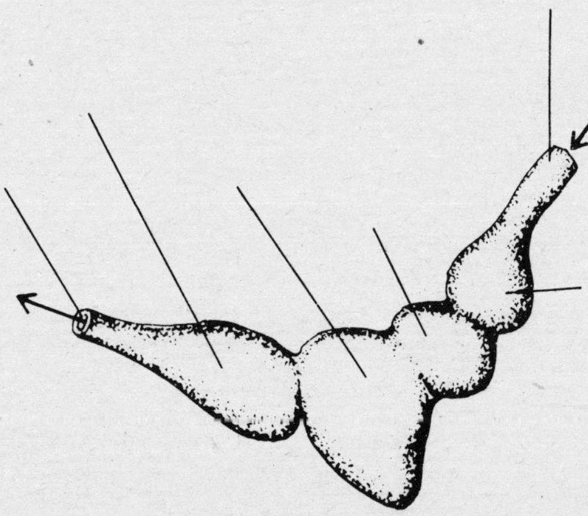

Fish heart

Summary

(1) What external structures have you observed that adapt a fish to life in water?

(2) Review the structures you have observed by completing the following diagram. Label: *gill filaments, liver, gall bladder, stomach, pylorus, pyloric caeca, intestine, anus, heart, gonad (testes* or *ovary), urinary bladder, urogenital opening, air bladder, kidney*.

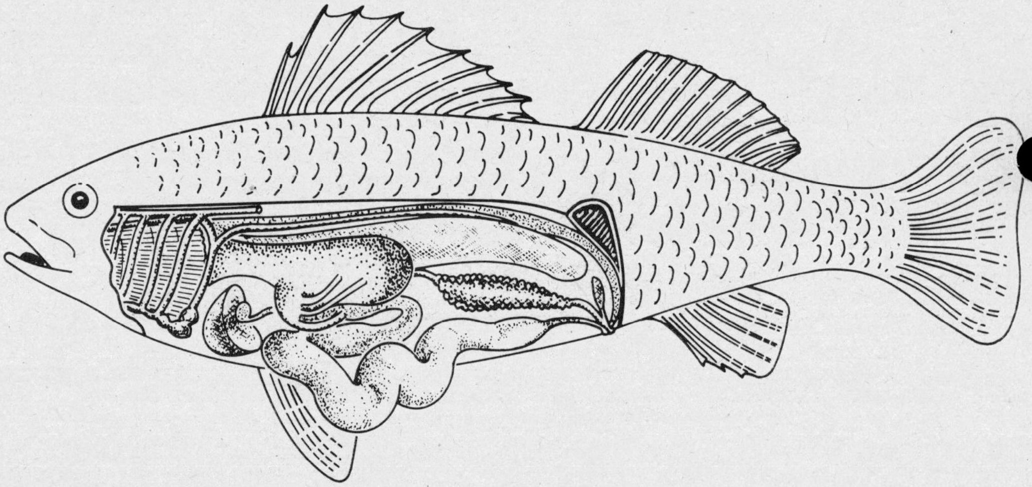

NAME DATE

Amphibian Vertebrates 35

MATERIALS

INVESTIGATION▶

- living adult bullfrog
- battery jar or aquarium
- freshly killed or preserved frog
- dissecting pan
- dissecting scissors
- probe
- forceps
- dissecting microscope

OBJECTIVES▶

- To observe how an adult frog is adapted for both aquatic and terrestrial habitats
- To observe organ systems of an amphibian vertebrate

Problem

What bullfrog adaptations enable it to live a "double-life"?

Salamanders, frogs, and toads are common amphibian vertebrates. They have bony endoskeletons and backbones. Several amphibian characteristics are: smooth, moist skin, two pairs of limbs with digits, upper and lower jaws, eyelids, no ears but membrane is present. They are adapted to water and land environments.

The life history of the bullfrog begins when eggs are laid and fertilized in a water environment. Tadpoles, adapted for an aquatic life, hatch from the eggs. The bullfrog tadpole spends two winters and two or three years in its various tadpole stages. As they feed and grow, they gradually develop adaptations for a life on land.

Part I. Analysis

The Class Amphibia includes three orders. Two orders are commonly seen. The Order Urodela includes salamanders. The Order Salientia includes frogs and toads. Observe several differences of these groups by comparing the drawings on the Data Sheet and answering question 1.

Part II. Investigation

(A) The adult bullfrog

Place a frog in a battery jar, aquarium, or sink with enough water so the animal can swim. Notice how the frog floats in the water. Touch the frog and watch it swim. Examine the eyes of the frog. Notice that the frog cannot move its eyes as you can. Move an object close to the frog's eye. Observe the thin membrane that covers the eye from below. This is the *nictitating membrane*. Record your observations in questions 2–5 on the Data Sheet.

Remove a frog and place it on a table under an inverted battery jar. Observe its breathing movements, paying special attention to the floor of the mouth and muscles of the body wall. Use your text to help you interpret your observations as you answer question 6 on the Data Sheet.

Notice the large eardrums *(tympanic membranes)* posterior to the eyes. These function to receive sound waves, which are transmitted to the inner ear mechanism. Turn to the first page of the Data Sheet and label the *tympanic membrane* on the frog drawing.

Examine the *forelegs* of the frog. Holding the animal carefully, stretch out the *hind legs*. Compare these two appendages and record your observations in question 7 on the Data Sheet.

Examine the coloration of the bullfrog on its dorsal and ventral surfaces. Consider how the coloration is adaptive and record your observations and conclusions in question 8 on the Data Sheet.

(B) Dissection of the bullfrog

Before starting your dissection, carefully read and be certain you understand all directions given in the procedure. Check the locations of the structures to be observed in the diagrams of frog anatomy in your textbook. The color, texture, and flexibility of organs will be different between preserved and freshly killed specimens.

Cut the *mouth* at each hinge joint to allow you to open the mouth wide. Notice the attachment of the *tongue*. In the living specimen the tongue is sticky to catch and hold insects.

At the back of the mouth are the openings to three structures. On a male frog, near the hinge of the lower jaw on either side, locate the opening to the *vocal sacs*.

Locate the slitlike *glottis* in the center of the rear of the mouth. Use a probe to determine where the glottis leads. Notice the wide *gullet* (opening of the esophagus) that allows the frog to swallow food whole.

Find the openings of the *Eustachian tubes* near the hinge of the upper jaw. Next, locate the *internal nostril openings* in the roof of the mouth. Remember, this position is an adaptation that allows the frog to breathe while partially submerged.

Notice two *vomerine teeth* extending from the roof of the mouth. Observe that only the upper jaw is provided with *small teeth*. Record your observations on the structures of the mouth by answering questions 9–11 on the Data Sheet. Then, label the mouth structures in the drawing on the Data Sheet.

Lay the frog on its dorsal side. Insert your scissors through the skin just above the anal opening and make incisions as shown in the diagram on the next page.

When you have made the incisions through the skin of the frog, fold the flaps of skin back, revealing the underlying muscles. Study the ventral and lateral muscle layers now exposed by the removal of skin. Muscles can only shorten when they contract.

Observe the large blood vessel lying along the midline under the muscle layer. This is the *abdominal vein*. It comes from the liver.

NAME DATE

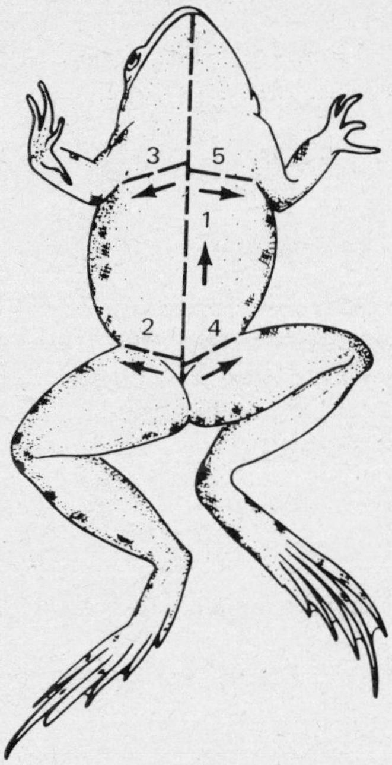

Repeat the same incisions as carried out on the skin, being careful not to damage underlying organs. Cut through the bones of the shoulder girdle and remove the muscle layers with forceps. The body cavity may be nearly filled with black-colored *eggs*. If so, your specimen is a female.

Many organs form the *digestive system*. Locate the large *liver* anterior in the body cavity. Identify its 5 lobes and find the greenish *gall bladder*. Notice the enlarged white tubular *stomach* under the liver. Observe that the anterior end of the stomach is larger than the *esophagus*. The stomach becomes smaller at the *pylorus* where it joins the *small intestine*. The *duodenum* is the short section of the small intestine leading from the pylorus. The highly coiled *ileum* is the major portion of the small intestine, which widens into the *large intestine*. Spread a portion of the coiled intestine apart and locate the membranous *mesentery*. Examine the mesentery closely and determine its function. Within the mesentery is a small, spherical *spleen*. The spleen functions as a blood storage organ. Answer question 12 on the Data Sheet.

The *pancreas* is attached to the lower part of the stomach and duodenum. Remove the liver, using care not to damage other organs. Cut through the gullet at its upper end. Raise the stomach and carefully cut the mesentery to free the intestine. The *cloaca* lies below the large intestine between the hind legs. Cut through the large intestine as close to the cloaca as possible. Label the parts in the diagram on the Data Sheet.

Using scissors, cut along the outer curvature of the stomach towards the pylorus. Pin the tissue of the stomach back and examine any contents that may be present. Note that the lining is ridged. Cut through the pylorus and note its highly muscular, ridged structure. Think about its function and answer question 13 on the Data Sheet.

Remove a short piece of the small intestine. Open and examine its lining with a dissecting microscope. Describe your observations in question 14 on the Data Sheet.

The *urogenital system* functions in reproduction and removal of wastes from the body. Locate the cream-colored *oviducts* in the female through which eggs pass from the anterior of the body cavity to the posterior opening, the *cloaca*. Remove the oviducts to reveal the brownish *kidneys*. The kidneys lie along the back on either side of the spine.

Notice that they are covered by a thin, tough membrane, the *peritoneum*, which lines the body cavity. Locate a small, twisting tube, the *ureter*, leading from each kidney to the cloaca. Attached to the cloaca is the *urinary bladder*. Label the parts in the diagram on the Data Sheet.

Do not confuse the ureters with oviducts in the female. If your frog is an immature female, you will find two lobed, grayish *ovaries* lying close to the kidneys. The testes of the male are in a corresponding position. Answer question 15 on the Data Sheet.

One vertebrate characteristic is a closed *circulatory system*. Various vertebrates have different heart structures. Examine the *heart* of your amphibian specimen. Notice that it lies in a thin sac, the *pericardium*. Use your probe to locate the great veins and arteries leading to and from the heart. (Consult your textbook to identify the larger blood vessels and chambers.) Examine the *lungs* and notice the *pulmonary veins* and *pulmonary arteries* connecting the heart and lungs. Remove the heart, leaving as much of the blood vessels attached as possible. Examine the front side under a dissecting microscope. Locate the *right atrium*, *left atrium*, and *ventricle*. The large vessel arising from the ventricle and forming a "Y" at the top of the heart is the *conus arteriosus*. This large vessel divides into the *left* and *right truncus arteriosus*. Three major arteries arise from each branch: the *carotid artery* to the head, the *aortic arch* to the body, and the *pulmocutaneous artery* to the lungs and skin. Examine the dorsal side of the heart. Three branches of the *vena cava* unite in a thin-walled sac, the *sinus venosus*, which opens into the right atrium. Record your observations on the heart by labeling the drawings on the Data Sheet.

Investigations On Your Own

(1) Observe the arrangement of the spots and the coloration of a leopard frog (use a preserved specimen). Remove a section of the dorsal skin containing one of the frog's spots and mount it on a slide in a drop of water, using a cover glass. Examine the entire section under low and high power of your microscope. Examine various sections, from pigmented spots of the dorsal surface to clearer areas of the ventral surface. Make drawings of your observations and write up your findings.

(2) Carefully dissect the brain of a frog. To do so, remove the skin from the dorsal surface of the head. The brain is encased within a bony cranial cavity. Remove the cranium by scraping away the bone with a scalpel. As you are able, break chips of bone away to reveal the brain. Expose it anteriorly, posteriorly, and to the sides as much as possible before attempting to remove it. To accompany your specimen, make a diagram showing the significant features of the frog's brain and label the parts.

NAME ____________________ DATE ____________

Amphibian Vertebrates 35

DATA SHEET

Analysis

Salamanders, frogs, and toads are common amphibian vertebrates. The Class Amphibia includes the Orders Urodela (salamanders) and Salientia (frogs and toads). Label each drawing with the Order. Then label: *anterior, posterior, dorsal, ventral.* Where applicable, label: *head, trunk, tail, 2 pairs equal limbs, fused head and trunk, 2 pairs of limbs unequal in size, webbed feet, large mouth*.

(1) What characteristics separate these two Amphibian orders? ____________________

____

____

____

Investigation

(A) The adult bullfrog

(2) Of what advantage is the floating behavior of the frog? ____________________

____

____

____

(3) Describe the swimming movements. ____________________

____

____

(4) Record your observations on the eye of the frog. Note the number of eyelids.

______

______

______

(5) Suggest an adaptive value to the nictitating membrane in water and on land.

______

(6) Differentiate mouth breathing from lung breathing in the bullfrog. ______

______

______

______

______

______

______

(7) Compare the digits, webbing, and length of the hindlegs and forelegs of the bullfrog.

______

______

(8) Why is the coloration of the frog considered to be protective? ______

______

______

______

(C) Dissection of the bullfrog.

(9) Describe the attachment of the tongue. ______

______

(10) Where does the glottis lead?

______

______

(11) Where do the Eustachian tubes lead? ______

______

Label Mouth Structures:

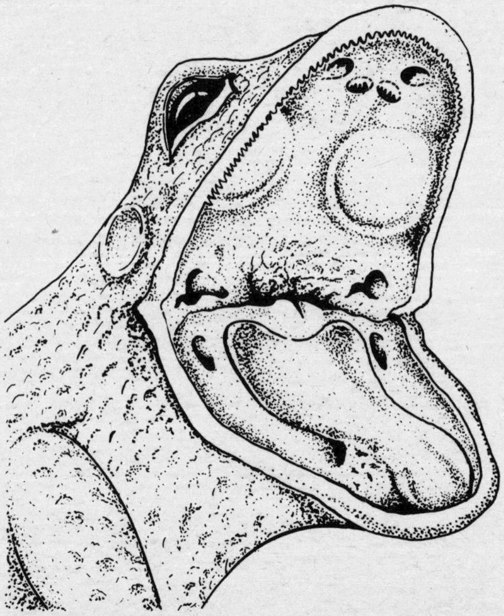

NAME DATE

(12) What is the function of the mesentery? ______

(13) What is the function of the pyloris? ______

(14) Describe the lining of the small intestine. ______

(15) What is the function of the urinary bladder? ______

Label the parts of the alimentary canal and urogenital system.

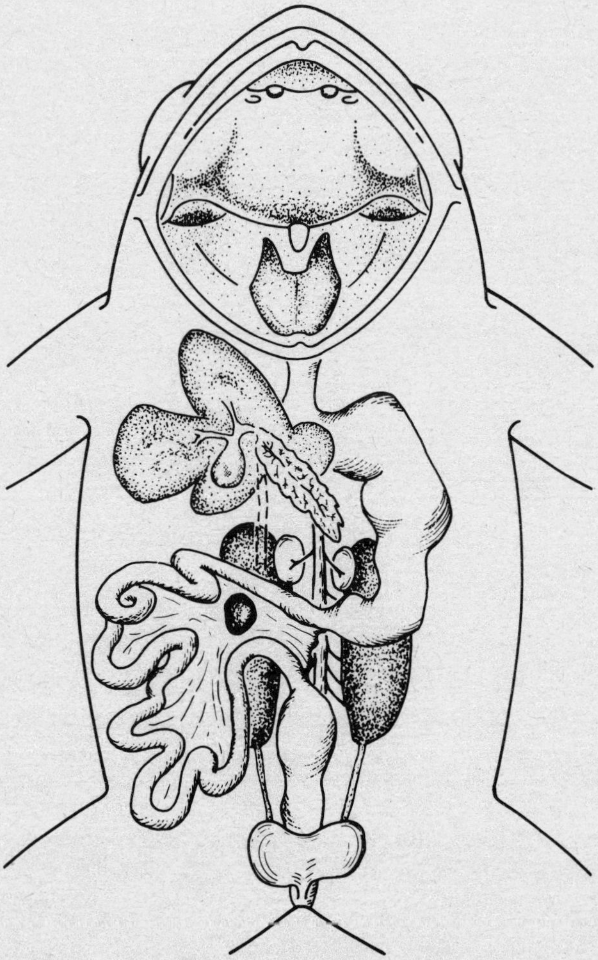

Label the parts of the heart.

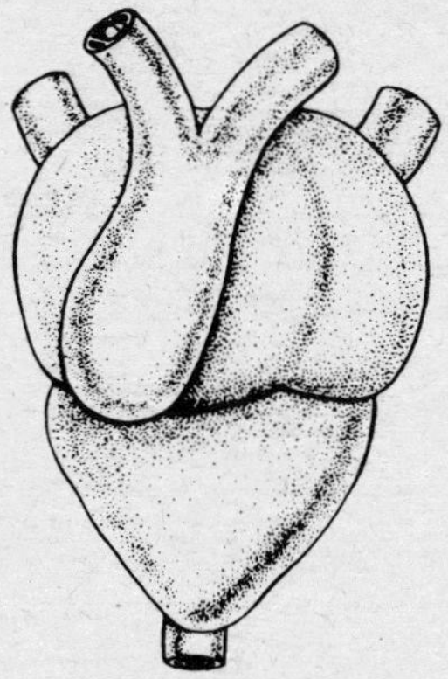

Front view

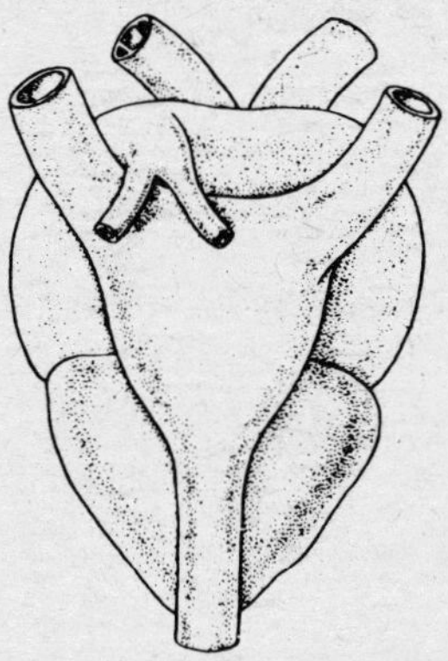

Back view

Summary

(1) Briefly relate how the mouth and its structures adapt a frog to its terrestrial existence.

(2) Of what advantage might it be for an animal, such as the frog, to have a completely different life in which the immature are aquatic and the adults both aquatic and terrestrial?

(3) Can you think of any disadvantages for such a "double life" of an amphibian?

NAME DATE

36 Reptilian Vertebrates

MATERIALS

INVESTIGATION▶

- live laboratory specimens of:
  turtle (charts will suffice)
  nonpoisonous snake
  lizard
- shed skin of any snake species
- hand lens
- dried turtle shell
- museum mount of snake skeleton

OBJECTIVES▶

- To identify characteristics of reptiles
- To study structural adaptations of turtles
- To observe external adaptations of snakes
- To compare structural features of snakes and lizards

Problem

What terrestrial adaptations can be observed in reptiles?

Internal fertilization and the amniote egg are adaptations for life on land. Many reptiles live in or around water, but many, such as snakes, lizards, and tortoises, can live in desert environments.

Two orders of the Phylum Reptilia are common and can be observed and compared to identify their characteristics. The Order Chelonia (Testudinata) includes turtles and tortoises. The Order Squamata includes snakes and lizards.

Part I. Analysis

Turn to the Data Sheet. Identify the order of each animal. Then answer question 1.

Compare the turtle, snake, and lizard by examining each one carefully. How many similar characteristics can you identify in the specimens? Record your observations in question 2 on the Data Sheet.

What obvious differences do you observe? Record your observations in question 3 on the Data Sheet.

Part II. Investigation

(A) The turtle

Observe the shell of the painted turtle. Notice that the head and appendages protrude from the shell. Touch the appendages and observe the turtle's reaction. Observe that the feet are webbed and have claws. Record your observations in question 4 on the Data Sheet.

Examine the lower shell or *plastron* and the upper shell or *carapace*. Notice that they are joined by a *bony bridge*. Record your observations in questions 5 and 6 on the Data Sheet.

Look at a dried turtle shell. Notice that the covering of the *horny shields* does not correspond with the pattern of the expanded ribs. Identify another skeletal feature fused to the underside of the carapace. Record your observation in question 7 on the Data Sheet.

Examine the head of the turtle. Notice that it has no teeth. Determine what it uses to bite its food. Observe the location of the nostrils. Examine the eyes and note the eyelids. Record your observations in questions 8–10 on the Data Sheet. Then, label the drawing of the turtle on top of the Data Sheet: *carapace, head, horny beak, nostrils, eyes, forelegs, hind legs, claws, webbing, tail*.

(B) The snake

Examine the snake closely and observe that the skin is covered with *scales*. Hold or touch the skin of the snake. Notice whether the snake's temperature appears to be more than, less than, or about the same as yours. Record your observations and conclusions in questions 11 and 12 on the Data Sheet.

Locate the *vent* or *cloacal opening*. Determine the 3 body regions of a snake and record your observation in question 13 on the Data Sheet.

Observe the scales with a hand lens. Notice how the scales are placed. Compare the scales of the living snake to those of the shed skin of any species of snake. Compare the ventral scales to those on the dorsal surface. Record your observations in questions 14 and 15 on the Data Sheet.

The ventral scales are known as *abdominal scutes*. They are internally attached by muscles to the ribs. Closely observe a snake from the side as it moves across the surface of a table or a rough board. Record your observations in questions 16 and 17 on the Data Sheet.

Examine the head of a snake. Examine its eyes for eyelids and ear openings. Record your observations in questions 18 and 19 on the Data Sheet.

Using a museum-mounted skeleton, examine the skull and *jaws*. Notice the structure of the upper and lower jaws. Notice how the teeth on the jaws are oriented to the rear of the mouth. Record your observations in questions 20–22 on the Data Sheet. Label the drawing of the snake on top of the Data Sheet: *head, trunk, tail, scales, scutes, vent, mouth, eyes*.

(C) The lizard

Examine the lizard closely and determine that it has a skin with horny epidermal scales. Observe the head of the lizard. Look for eyelids and ear openings, and record your findings in questions 23 and 24 on the Data Sheet.

Observe the feet of the lizard. Look for webbing between the toes and claws on the ends of the toes.

Investigations On Your Own

Dissect a nonpoisonous snake to observe the organ-systems. Make diagrams and label the structures you find. When you have completed your dissection, you may use thread to attach it to a piece of glass, cut to fit a museum jar. When the jar is filled with a preservative, a plastic cover will preserve your dissection for use by other students.

NAME ______________________ DATE ______________

36

Reptilian Vertebrates

DATA SHEET

Analysis

Write the name of the order of the turtle, snake, and lizard in the drawings below. On each drawing identify: *anterior, posterior, dorsal,* and *ventral*.

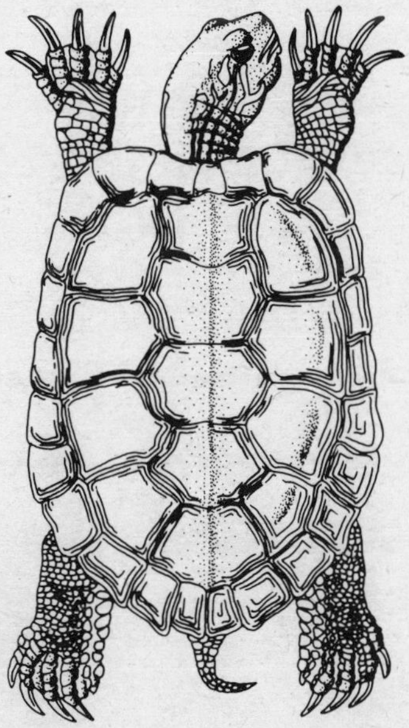

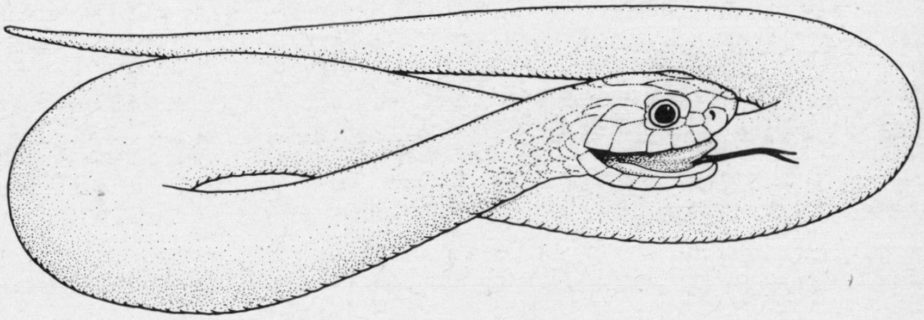

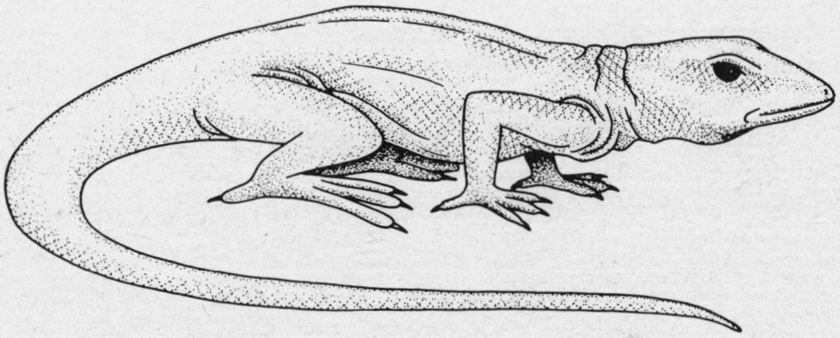

(1) What two important adaptations for life on land cannot be seen in the above drawings of reptiles? ______________________

(2) What observed characteristics are shared by the turtle, snake, and lizard? ______________________

(3) What obvious differences have you observed?

turtle: ______________________

snake: ______________________

lizard: ______________________

(A) The turtle

(4) How is the reaction of the turtles to touch an adaptation? ______________________

(5) Compare the plastron and the carapace of the painted turtle. ______________________

____

____

____

____

(6) Explain how the shell color of the painted turtle may be adaptive. ______________

____

____

(7) What skeletal feature is fused to the underside of the carapace? ______________

____

(8) What structure is used by the turtle to bite its food? ______________________

____

(9) How is the location of nostrils considered an adaptation for the painted turtle?

____

____

(10) Record your observations of the turtle's eyelids. ______________________

____

(B) The snake

(11) How does the snake's skin adapt it to land? ______________________

____

____

(12) Describe the feeling of the snake's skin. ______________________

____

(13) What structures determine the 3 body regions of the snake? ______________

____

____

____

____

(14) Describe the placement of scales on the living snake. ______

(15) Compare the ventral and dorsal scales of a snake. ______

(16) Describe the snake's lateral undulatory movement. ______

(17) How are the scutes adapted for locomotion? ______

(18) Are eyelids present on the snake? ______

(19) Are ear openings present on the snake? ______

(20) What difference can be seen in the structure of the upper and lower jaws? ______

(21) On a living specimen, what holds the lower jaws together? ______

(22) What structural adaptation of the jaw and gullet permits a snake to swallow prey four to five times larger than its throat? ______

(C) The lizard

(23) Explain how you reached your conclusion regarding eyelids on the lizard. ______

(24) Does the lizard have external ear openings? ______

Summary

(1) State 6 characteristics of the Phylum Reptilia that adapt the animals to land.

(a) ______

(b) ______

(c) ______

(d) ______

(e) ______

(f) ______

(2) How do turtles differ from snakes and lizards? ______

(3) What characteristics separate snakes from lizards? ______

(4) It is believed that turtles have been on earth longer than any of the present-day reptiles. Give several reasons why they have survived so well. ______

NAME DATE

Avian Vertebrates 37

MATERIALS

INVESTIGATION▶

- Museum mounts (or drawings) of the following:
- hawk, owl, woodpecker
- loon, quail, wood thrush
- pelican, hummingbird
- heron, dove, duck
- kingfisher

OBJECTIVES▶

- To observe characteristics of birds
- To identify avian adaptations for various environments

Problem

What characteristics are found in all birds, and how are various structures modified to adapt birds to various environments?

The streamlined, feather-covered body of birds adapt them for rapid movement. The forelimbs of birds are specialized as wings, which most birds use for flight. The hindlimbs support the bird's body when the animal is not flying. Light, porous bones also adapt birds for flying.

The four-chambered heart of birds allows for more efficient circulation and oxygen distribution through their warm-blooded bodies. Birds are widely distributed because the amniote egg, encased in a lime-containing shell, adapts them for a life on land.

Procedure

Part I. Analysis

Turn to the Data Sheet. Using your text to help you, answer questions 1–4 to identify several characteristics common to all birds.

Part II. Investigation

(A) The beaks of birds

The mouth of a bird is formed by a toothless, horny beak. Look at the many different shapes of beaks in the specimens. These are adaptations for eating many different kinds of food from the environment. Consider the beak structure and function of each.

stout and short — eat small seeds
stout and spear-shaped — spearing fish
chisel-shaped — drilling for insects
hooked — to catch prey
tubular — to suck nectar
long and stout — to scoop fish
short — multipurpose
flat and somewhat hooked — straining algae and small organisms

When you have determined the beak modification of each specimen, record your conclusions in the chart on the Data Sheet. Then answer questions 5–7 on the Data Sheet.

(B) The feet and legs of birds

Observe the legs of the specimens and notice that they are covered with scales. Look at the arrangement of the toes on the feet. Consider the habitat of each specimen and determine how the legs and feet are adaptations. Record your conclusions in the chart on the Data Sheet and then, in questions 8–14, identify each bird by the placement of its toes.

Quail
Woodpecker
Owl
Thrush
Hummingbird
Hawk
Pelican
Loon
Heron
Dove
Duck
Kingfisher

NAME ______________________________ DATE ______________

Avian Vertebrates 37

DATA SHEET

Analysis

Using your text to help you, identify several characteristics of the Class Aves by answering questions 1–4.

(1) How do the shape and the body covering adapt the bird to flight? ______________

____

____

(2) What avian adaptations are seen in the limbs of birds? ______________

____

____

(3) How is the structure of the heart of a bird adaptive? ______________

____

____

____

(4) Why is the bird's egg considered to be an adaptation? ______________

____

____

Investigation

(A) The beaks of birds

(5) For what are the beaks of the loon and the heron adapted? ______________

____

(6) The heron and the pelican are both fishing birds. What do their beaks indicate about their feeding behavior? ______________

____

____

____

(7) Which birds have beaks adapted for tearing flesh? ______________________

(B) The feet and legs of birds
In the following descriptions, the numbers refer to the placement of the toes. Identify each specimen, and then state the function of the particular arrangement of toes.

(8) 2 in front, 2 behind ______________________

(9) 3 in front, 1 behind, long legs ______________________

(10) 3 in front, 1 behind, webbed ______________________

(11) 3 in front, 1 behind, large with talons ______________________

(12) 3 in front, 1 behind, legs medium length ______________________

____

(13) 3 in front, 1 behind, legs short ______________________

(14) 3 in front, 1 behind, legs very short and weak ______________________

____

Summary

(1) What is a bird? ______________________

____

____

____

____

____

____

____

____

(2) Birds and reptiles are both adapted for life on land. What similarities are found in both these vertebrate groups? ______________________

____

____

(3) The loon, heron, and kingfisher all have the same beak adaptation for spearing fish. What does their leg structure indicate about their feeding habits? ______________________

____

____

NAME DATE

Name of Bird	Habitat	Beak Adaptation	Feet Adaptation
1. Woodpecker			
2. Quail			
3. Thrush			
4. Hawk			
5. Owl			
6. Loon			
7. Pelican			
8. Hummingbird			
9. Heron			
10. Dove			
11. Duck			
12. Kingfisher			

Investigations On Your Own

(1) Saw a chicken's leg bone (femur) in half. Make a drawing and label the adaptations for flight. Compare the bone to a mammal's femur.
(2) Carefully dissect a bird's brain. Make and label drawings and preserve the brain for your school museum. Compare the brain of the bird to that of a reptilian vertebrate. Write a paper on your findings.
(3) Carefully dissect a bird's heart. Make and label drawings and preserve the heart for the school museum. Compare the heart of the bird to that of a reptile. Write a paper on your findings.
(4) Collect, observe, and draw the several kinds of feathers found on birds. Explain the function of each.

NAME DATE

The Fetal Pig, A Mammalian Vertebrate 38

MATERIALS

INVESTIGATION▶

- preserved fetal pig
- dissecting pan or tray
- dissecting scissors
- scissors
- forceps
- probe
- dissecting needles

OBJECTIVES▶

- To dissect a fetal pig
- To identify the major organs of the fetal pig
- To understand characteristics of fetal mammalian life

Problem

What mammalian characteristics can be observed in a fetal pig?
Specialization in an organism is an indication that an adaptation has occurred. This specialization helps the organism adapt to its environment. Differences in the structure of mammals have led investigators to believe that mammals have the highest degree of specialization because of their highly developed organ systems.

As the mammalian organ systems develop, important changes occur. Some of these involve the circulatory and respiratory systems. These can be observed in the fetal pig.

Procedure

Part I. Analysis

Look at the diagrams of the mammalian heart and brain on the Data Sheet. Label only those parts that show characteristics in mammals that are more highly advanced and complex than other animals. Answer questions 1–3.

Part II. Investigation

(A) External anatomy
Obtain a fetal pig and place it in a dissecting pan. Examine the *umbilical cord.* Observe that it contains three blood vessels. Identify *one large vein* and *two smaller arteries.* Answer questions 4 and 5 on the Data Sheet. The diagram in the *Summary* of the fetal pig dissection can be labeled by students as they proceed with the dissection.

Examine the *ears.* With a probe, locate the eardrum. Notice that the eyes have an upper and lower lid and a small mass of tissue in the upper corner (the nictitating membrane). This membrane has no function in the pig.

Examine the *feet.* Notice that the first digit of both the fore and hind limb is absent and the second and fifth are reduced in size but remain functional. Answer question 6 on the Data Sheet.

(B) The digestive system
In the dissection and observations of the internal organs, you will proceed by systems and *remove organs only when directed to do so.* Study and use diagrams to aid in your observations of the internal organs. As you dissect, keep in mind the interrelationships of systems. While concentrating on a single system, use care not to damage other systems.

Use two pieces of strong twine and tie one around a wrist and one around an ankle of the pig. Pull each under the dissecting pan and tie the twine tightly to the opposite wrist or ankle. To open the abdominal cavity, make incisions as indicated in the diagram on the next page.

Fetal pig ventral external view

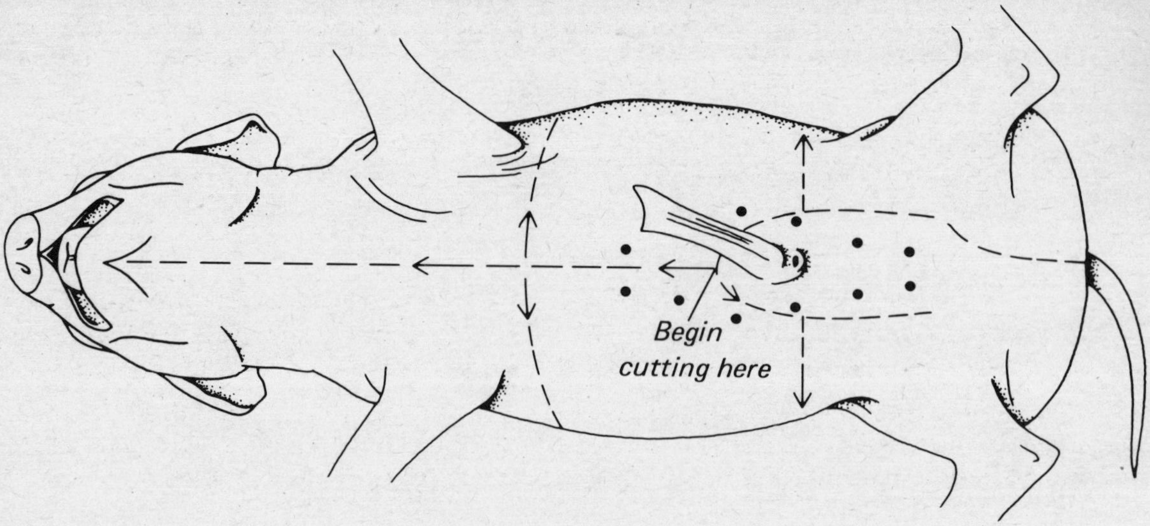

Cut carefully with scissors to avoid damaging the underlying organs. Pull the skin and muscle layers aside to expose the *peritoneum,* the lining of the abdominal cavity. Locate the *umbilical vein.* Lay back the cord and strip it of its skin. The large, brownish-red organ is the *liver.* Lying beneath and to the right of the liver is the cream-white *stomach.* Answer questions 7 and 8 on the Data Sheet.

Locate where the *esophagus* joins the stomach and where the *small intestine* leaves. The first 3–4 centimeters of the small intestine is the *duodenum.* Observe that the small intestine is not loose in the abdominal cavity but is held in place by the *mesentery.* The *large intestine* appears as a compact coil. Locate the junction of the small and large intestines. Below this junction may be found a large pouchlike structure, the *caecum.* The *rectum* lies in the peritoneum of the dorsal body wall. Cut along the outer curvature of the stomach and observe the *rugae.* These are ridges within the stomach. Where the stomach joins the small intestine, observe a sphincter muscle known as the *pyloric valve.*

Several organs of the abdominal cavity are associated with the process of digestion but are not part of the digestive tract. Observe the liver more closely on its underside. Notice the *hepatic ducts* leading from the liver. They connect to the *cystic duct,* which extends to the *gall bladder.* These ducts join to become the *common bile duct.*

The *pancreas* is a large, white, granular organ located below the stomach. The *pancreatic duct* is difficult to find in the fetal pig. It enters the duodenum posterior to the entrance of the bile duct. The red, elongated organ extending around the outer curvature of the stomach is the *spleen.* Answer questions 9–18 on the Data Sheet.

Remove the organs of the digestive system that you have observed. Using your textbook, determine the function(s) each organ has in the process of digestion and write them in the table on the Data Sheet.

(C) The thoracic cavity

Fully expose the organs of the thoracic cavity by carefully cutting the flaps of tissue formed by your initial incision. Examine the placement of organs. Answer question 19 on the Data Sheet.

Notice the muscular sheet that separates the thoracic cavity from the abdominal cavity. This is the *diaphragm.* Answer question 20 on the Data Sheet.

Locate the large blood vessel, the *posterior vena cava,* which passes through the liver and through the heart. Carefully clip the edges of the diaphragm and separate the diaphram to drop it down over the liver.

Observe that the *heart* lies in a thin sac, the *pericardium,* and is partially covered by the thymus gland. (DO NOT REMOVE THE HEART AT THIS TIME!)

Locate the spongy *lungs* on either side of the heart. They appear collapsed because they have never been filled with air. The *trachea* in the neck region is easily identified by its cartilaginous rings. The rings hold the trachea constantly open to prevent its collapse.

With forceps, carefully pick away some of the connective tissue holding the heart and lungs in place. Observe where the trachea divides to form an inverted Y just above the lungs. Below the trachea in the neck and posterior to where the trachea divides to enter the lungs is the gullet or esophagus.

Continue the dissection into the neck, following the trachea. Note that the thymus gland extends far into the neck. Also in the neck are the *lymph glands.* Probe between the lobes of the lymph glands and locate the reddish, bi-lobed *thyroid gland* lying over the trachea. The enlargement at the anterior end of the trachea is the *larynx* or voice box, which contains the vocal cords. Leave the lungs and other organs in place.

Examine the mouth cavity containing the *tongue* and *teeth.* Notice the upper *hard* and *soft palate* forming the roof of the mouth. To the posterior of the mouth is the *pharynx,* a common passage for food going to the esophagus and air going to the lungs. To demonstrate the latter, run a needle probe through a nostril. Observe that the needle appears in the pharynx.

Locate the *epiglottis,* a flaplike structure at the top of the larynx. Answer question 21 on the Data Sheet.

Determine the function of the organs you have observed in the thorax, neck, and head regions. Record the system to which each organ belongs and its function in the chart on the Data Sheet.

(D) The circulatory system

To observe the heart, remove the *pericardium* and observe the ventricles. Notice the larger left ventricle and observe that the *right atrium* and *left atrium* are about the same size. Touch the atria and the ventricles with your finger. Record your observations in question 22 on the Data Sheet. Locate the *anterior* and *posterior vena cava.* These carry blood from the anterior and posterior portions of the body, respectively. Locate where they enter the heart. Find the *pulmonary veins* that carry blood from the lungs. Determine the chamber from which the veins arise. Answer questions 23–25 on the Data Sheet.

A characteristic feature of the fetal mammalian heart is the *ductus arteriosus.* This short vessel allows blood to bypass pulmonary circulation until birth, at which time there is a complete closure of the vessel. Answer question 26 on the Data Sheet.

To study the heart in further detail, it must be removed. Removal is accomplished by severing blood vessels that connect to it. Begin with the posterior vessels and progress anteriorly. In the following order, locate and cut: the *posterior vena cava, pulmonary veins, anterior vena cava, aorta,* and *pulmonary arteries.* On the surface of the heart are the *coronary arteries* and *veins.* Then answer questions 27 and 28 on the Data Sheet.

To open the heart, make a slit along the midline on its ventral side. (This is the side you normally observe.) Identify the four chambers. Use your textbook to identify the internal features that you are able to observe. Locate the entrance and exit of the veins and arteries connected to the heart, as well as the valves that prevent backflow into the heart chambers. Compare the walls of the atria and ventricles.

(E) The urogenital system

The urinary or excretory system and genital system are structurally related. Therefore, it is convenient to study them together. Recall that you are dealing with paired structures.

Using forceps, carefully remove any fat that prevents easy observation of the urogenital organs. On the dorsal side of the abdominal cavity, outside the peritoneum, are the prominent bean-shaped *kidneys*. These organs function to remove the wastes of metabolism carried to them by the blood.

Locate the *ureter* originating from the concave side of the kidney. Follow the ureter posteriorly until it joins the *urinary bladder*. Do not remove these organs.

Prepare for the observation of the reproductive organs of the male or female by pulling the hind legs apart. With scissors, cut anteriorly a little to one side of the mid-ventral line to avoid cutting the penis on the male. Press firmly on the tissue between the legs to feel the cartilaginous structure of the *pubic symphysis*. This is part of the pelvic girdle. Continue the incision anteriorly and cut through the pubic symphysis. Expose the *urethra*. Answer question 29 on the Data Sheet.

The Male Reproductive System. Examine the *scrotal sacs* at the posterior end of the male pig. Open one sac and determine the presence of a *testis*. If your specimen is advanced in fetal development, the testes may have already descended into the scrotal sacs. Otherwise, they may be found in a tubelike structure, the *inguinal canal*, as small, oval organs. In either case, locate one of the testes and note the coiled tubule making up the *epididymis*. Follow this tube forward as it passes through the inguinal canal as the *vas deferens*. Follow the vas deferens. Note how it loops over the *ureters* and enters the *urethra*. At this point, as in humans, the urethra becomes a *urogenital duct*. It carries urine as well as reproductive products.

Follow the urethra until it passes the *penis*. The penis lies in a sheath of tissue in the ventral abdominal wall. The urogenital opening is just posterior to the umbilical cord. Answer question 30 on the Data Sheet.

The Female Reproductive System. Exchange specimens to have the opportunity to examine both sexes. Spread the legs to separate the pubic symphysis and thereby expose the female reproductive system. Locate the oval-shaped *ovaries*, which are found posterior to the kidneys. Leading from the ovaries are the twisted *oviducts* or *Fallopian tubes*. The oviducts continue posteriorly and are soon supported by broad *ligaments*. Further on, the oviducts join to form the *common uterus*. You will notice a slight constriction of the common uterus marking the location of the *cervix*. Posterior to the cervix, the remainder of the tube forms the *vagina*. Locate the point where the urethra joins to form the *common urogenital sinus*. Answer question 31 on the Data Sheet.

NAME _______________________________ DATE _______________

The Fetal Pig, A Mammalian Vertebrate 38

DATA SHEET

Analysis

On the drawings of the mammalian heart and brain, label those parts that are more highly advanced.

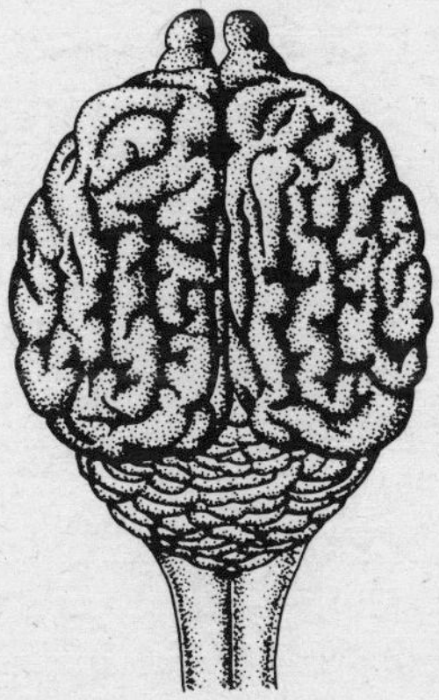

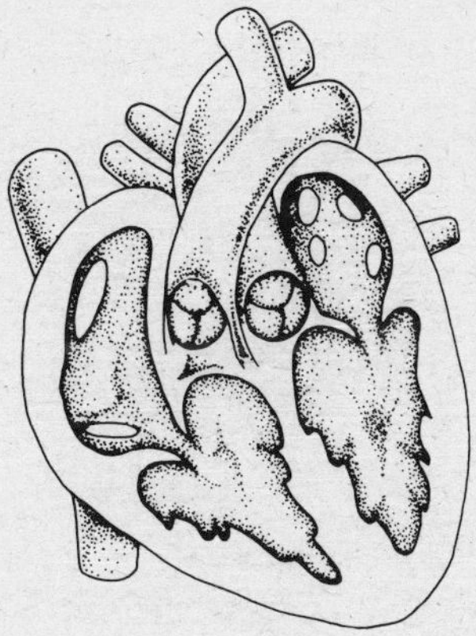

(1) What is the basic difference between a bird heart and a mammalian heart? ______

____

(2) What is the advantage of the heart as a double pump? ______________

____

(3) What does the structure of the mammalian brain indicate about intelligence? ______

____

____

Investigation

(A) External anatomy of the fetal pig

(4) What differences did you observe in the structure of an artery and vein of the umbilical cord? ______________

____

(5) How does a fetus get rid of its waste products? ______________

____

____

(6) How many digits are present? ______________

(B) The digestive system

(7) To what organ does the umbilical vein lead? ______

(8) How many lobes does the liver have? ______

(9) Describe the mesentery. ______

______

(10) Where is the mesentery attached? ______

(11) In humans, what structure is found at the junction of the small and large intestine?

______

(12) What is the posterior opening of the digestive tract called? ______

(13) How do the rugae within the stomach aid in mechanical digestion? ______

______

______

(14) What is the function of the pyloric valve? ______

______

(15) What saclike structure is located on the underside of the liver? ______

______

(16) Describe its appearance. ______

(17) Where does the common bile duct lead? ______

(18) Of what system is the spleen functionally a part? ______

______

Organ	Function
stomach	
duodenum	
small intestine	
large intestine	
liver	
gall bladder	
pancreas	

(C) The thoracic cavity

(19) What essential organs are in the thoracic cavity? ______

NAME DATE

(20) Does the diaphragm function in the fetus? Explain your answer.

(21) What is the function of the epiglottis?

Organ	System	Function
thymus gland		
lungs		
trachea		
thyroid gland		
lymph glands		
esophagus		

(D) The circulatory system

(22) What differences between the atria and ventricles could you feel with your fingers?

(23) Into what chamber of the heart do the anterior and posterior vena cava open?

(24) From what chamber does the aorta arise?

(25) To what structures do the pulmonary arteries lead?

(26) Why does the ductus arteriosus close off at the time of birth?

(27) What is the function of the coronary circulation?

(28) What results when coronary circulation is prevented?

(E) The urogenital system

(29) Name the structures (in order) through which urine passes from the kidney of the pig. ____________________

(30) Name the structures (in order) through which sperms would pass from the testes of the pig. ____________________

____

(31) Name the structures (in order) through which an egg would pass from the ovary of the pig. ____________________

____

Summary

Test yourself by labeling the following diagram:

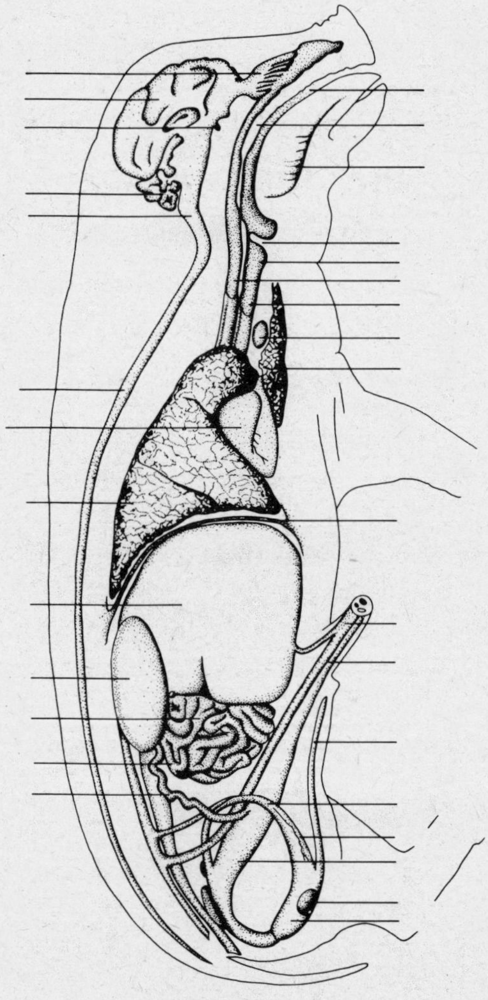

Internal anatomy of male, right side

NAME DATE

Muscular Activity 40

MATERIALS

ANALYSIS▶
- charts of muscles

INVESTIGATION▶
- prepared slides of smooth muscle, cardiac muscle, and skeletal muscle
- microscope
- clock or watch with a second hand

OBJECTIVES▶
- To understand that skeletal muscles work in pairs
- To identify smooth, cardiac, and skeletal muscle tissues
- To measure some changes that take place during muscular activity

Problem

Are there different kinds of muscle tissue? What happens during muscular activity? Muscular tissue has the ability to contract and to produce movement. The three kinds of muscle tissue are: smooth, cardiac, and skeletal. Skeletal muscle tissue is made up of long, tapered cells with striations or dark bands. Skeletal muscles are the voluntary muscles, which can be controlled at will. Cardiac muscles are also striated but they differ from skeletal muscles by their branching network of cells. Cardiac muscle is the involuntary muscle found in the heart. Smooth muscles are made of spindle-like cells with nuclei (one per cell) near the center. They are not striated. Smooth muscles are found in the internal organs of the stomach and intestines.

During muscular activity, muscle cells contract. The energy for muscular contraction comes from ATP, activated by nerve impulses. Skeletal muscles work in pairs — as one contracts, the other relaxes.

Procedure
Part I. Analysis

Make some basic observations about muscle activity by testing yourself. While sitting quietly, rest your right arm on your desk with the palm of your hand up. Place your left hand lightly around your right arm as shown in the figure at the right.

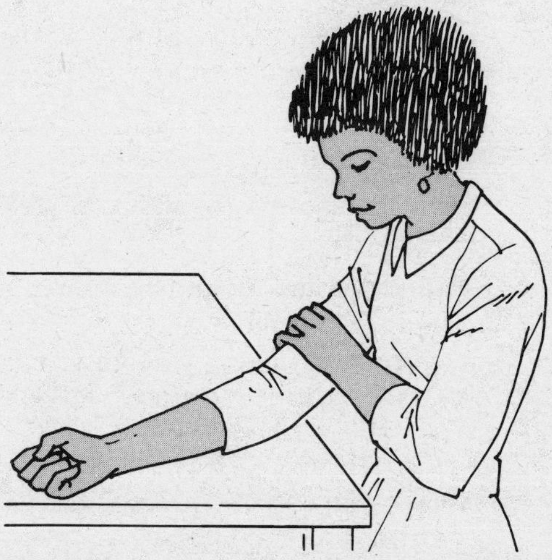

Make a fist and tightly bend (flex) your right arm. Make a note of the changes you feel by answering Data Sheet question 1.

Now fully straighten (extend) your right arm. Record any changes and observations by answering questions 2–3 on your Data Sheet.

With the aid of charts or your text, identify the *flexor muscle* and the *extensor muscle* you used in this activity. Note where these muscles are attached and the pivotal joint where movement takes place. Then answer questions 4–9.

It is important to remember that all body cells require energy to do their work. ATP is the energy of these active muscle cells. Recall what happens to the ATP molecule as energy is used (see Investigation 6). Answer question 10.

Investigation

(A) Carefully examine the three kinds of muscle cells: smooth, cardiac, and skeletal. Use the low power objective to find differences among the three muscle tissue samples. Sketch each tissue in the spaces provided on your Data Sheet. Label the nuclei, cytoplasm, and cell membrane in the smooth tissue; nuclei, cytoplasm, and striations in the cardiac and skeletal tissues. Then complete questions 11–16.

(B) The speed at which nutrients reach the muscles is indicated by the rate of the heartbeat. You can easily measure this by taking your pulse. Hold the fingers of one hand on the wrist of the other as in the drawing below. Count your pulse for 30 seconds. Double the number to find your pulse rate per minute. Another way to find your pulse rate is to *gently* place your right index finger and your middle finger on the side of your neck. Once again, count the "throbs" for 30 seconds and double your number. See the drawing below.

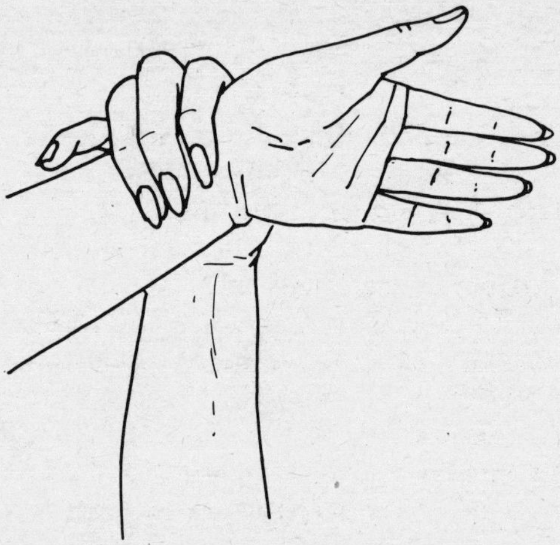

The rate of breathing is another measurement that indicates the oxygen requirements of muscles. Have your laboratory partner measure your respiration rate for one minute. (This is done by counting the rise and fall of the chest.) Record the "sitting position" respiration rate in the Data Sheet Table.

Now stand up and walk in place for one minute. In the Table, record your respiration rate and your pulse rate immediately. Wait 2 minutes and record these measurements again. If necessary, repeat the measurements until the rates are back to your "sitting position" rate.

Stand up again and run in place vigorously for exactly one minute. Sit down and immediately make the pulse and respiration rate measurements. Repeat these measurements every two minutes until the "sitting position" rates are reached. Record your results in the Data Sheet Table and complete questions 17–23.

Investigations On Your Own

(1) Many physical fitness tests include measurements like those you have made in this Investigation. Locate two or three such tests from your physical education teacher or the library. Compare these tests and explain how the measurements they require indicate "fitness."

(2) Explain how regular exercise over a period of time might change the results obtained in this Investigation. Can you devise an experiment that might prove your hypothesis?

NAME _______________________________ DATE _______________

Muscular Activity 40

DATA SHEET

Analysis

Follow the instructions for testing muscular activity in your arm as detailed on your Procedure page.

(1) Explain the changes you feel in your right arm as you make a fist and flex your arm. _______________

(2) Describe the changes you feel as you straighten (extend) your arm. _______________

(3) What observations show that skeletal muscles work in pairs? _______________

(4) What is the name of the *flexor muscle* involved in this activity? _______________

(5) Name the *extensor muscle*. _______________

(6) Where are these muscles attached? _______________

(7) At what joint does movement take place? _______________

(8) What must occur in the cells of the biceps when the arm is flexed? _______________

(9) What must occur in the cells of the triceps when the arm is flexed? _______________

(10) What chemical molecule supplies the energy for muscle activity? _______________

Investigation

(A) Sketch the following muscle tissues: smooth, cardiac, and skeletal. Label the *nuclei, cytoplasm*, and the *plasma membrane* in the smooth muscle tissue. Label the *nuclei, cytoplasm,* and the *striations* in the skeletal muscle tissue.

smooth muscle tissue

skele musc tissue

cardiac muscle tissue

(11) How does cardiac muscle tissue differ from skeletal muscle tissue? ______________

____

____

____

(12) Name one similarity in cardiac and skeletal muscle tissues. ______________

____

(13) In what way is smooth muscle tissue distinctive? ______________

____

____

(14) Name one place in the body where you would find cardiac muscle. ______________

smooth muscle. ______________________________

skeletal muscle. ______________________________

(15) Which type of muscle tissue is involved in bodily movements? ______________

(16) Which kinds of muscle tissues are important in involuntary activities? ______________

____

NAME DATE

(B) Measure your pulse rate using one of the methods described on your Procedure page. Then measure your respiration rate. Repeat the measurement of pulse and respiration rates after walking and then after running. See the Procedure page for specifics.

(17) What is your pulse rate per minute? ______

______

(18) What other system (besides circulatory) is involved in the heartbeat rate? ______

______

(19) What is your respiration rate? ______

Record your pulse and respiration rates under the conditions stated in the following chart: sitting position, walking, and running.

		Immediately	2 min.	4 min.	6 min.	8 min.
SITTING POSITION	pulse rate					
	respiration rate					
WALKING	pulse rate					
	respiration rate					
RUNNING	pulse rate					
	respiration rate					

(20) When taking your pulse after running in place, did you observe anything about the *strength*? Explain. ______

______

(21) How long did it take to "recover" from walking in place? ______

______

(22) How long did it take to "recover" from running in place? ______

______

(23) Explain why your answers to **(21)** and **(22)** are different. ______

______

Summary

(1) Explain the statement: "Skeletal muscles that move joints work in opposite pairs."

___

___

___

(2) Why was it necessary to take the "sitting position" measurements first? ______

___

___

Test what you have learned by filling in the blanks in the following sentences. Select answers from the word list.

Word List

ATP

biceps

bends

carbon dioxide

voluntary

oxygen

straightens

triceps

(3) An extensor muscle ______ a joint.

(4) A flexor muscle ______ a joint.

(5) Skeletal muscles are under your control and are said to be ______.

(6)–(7) Increasing your muscular activity will also increase the muscle's need for ______ and ______.

(8) A waste product given off during muscular activity is ______.

(9) The contraction of the ______ causes the elbow to bend.

(10) The contraction of the ______ straightens the arm at the elbow joint.

(11) How does this study of muscular activity indicate the coordination of several systems in your body? ______

___

___

___

NAME DATE

Starch Digestion 41

MATERIALS

ANALYSIS▶
- tracing paper, scissors, construction paper

INVESTIGATION▶
- cooked starch solution
- paraffin
- cheesecloth
- 25-mL beaker
- 3 test tubes
- glass-marking pencil
- microscope slide
- iodine solution
- 250-mL beakers
- warm water bath (37°C)
- Benedict's solution
- cellulose dialysis tubing (3–6 cm in diameter and 30 cm long)
- 5% glucose solution
- 200-mL beakers
- short piece of dowel or glass rod

OBJECTIVES▶
- To understand the action of enzymes by using molecular models
- To observe the activity of a digestive enzyme
- To determine why starches must be chemically changed before they can be absorbed

Problem

How are starches chemically changed in digestion? How are starches absorbed by the body?

Food supplies us with the energy we need to live. But before complex food molecules can be absorbed and then used, they must first be broken down (by digestion) into smaller, simpler molecules. Part of the digestive or breaking-down process is accomplished by means of chemicals called enzymes.

In starch digestion, the enzyme *amylase* converts a starch molecule into maltose, and finally into glucose. (Glucose is then used in the cellular respiration cycle with the energy stored in ATP.)

Procedure

Part I. Analysis

Trace and cut out 3 glucose models and 1 model of the enzyme amylase from the drawings on your Data Sheet page. Attach the glucose models to one another to represent a small part of the starch molecule chain. Note what must be lost before the glucose models can be attached.

Fit the amylase model between the last two glucose models. The enzyme amylase *weakens* the bond and causes a glucose molecule to separate from the starch molecule chain. (Enzymes are catalysts — they cause chemical changes but they do not change themselves.)

Now place the enzyme amylase model between the remaining glucose models. Once again a glucose will separate from the starch chain. Note what remains of the complex starch molecule. Answer Data Sheet questions 1–4.

Part II. Investigation

(A) In Analysis, you worked with models to simulate the chemical digestion of starch. Now you will test what happens during the digestion of starch, by experimenting in the laboratory.

Rinse out your mouth with water. Chew a piece of paraffin to stimulate the flow of saliva. Collect about 15 mL of saliva in a 25-mL beaker (covered with several layers of cheesecloth to serve as a filter). With a glass-marking pencil, divide a microscope slide into 2 sections. Place a drop of saliva in one section and a drop of cooked starch solution in the other section. Test each with a drop of iodine solution. Note your results by answering questions 5–6. (A blue-black color is a positive test for starch.)

Number 3 test tubes. Into tube *1*, put 3 mL of saliva. Into tube *2*, put 3 mL of cooked starch solution. Into tube *3*, put 3 mL of saliva and 3 mL of cooked starch solution. Place all 3 test tubes in a warm-water bath and then put them in the water bath. Observe any color changes that occur. (A yellow, green, or red color indicates the presence of a simple sugar, such as glucose.) Answer Data Sheet questions 7–9.

(B) Soften the dialysis tubing by soaking it in a beaker of water for half an hour. Measure 100 ml of water into each of 3 beakers and set them on tripods over Bunsen burners. Prepare the warm-water bath set-up as in the drawing below.

Tie the ends of 3 dialysis tubings with string, and put 10 mL of 5% glucose solution in the first; 10 mL of cooked starch solution in the second; and 5 mL of saliva plus 5 mL of cooked starch solution in the third.

Fasten the ends of the loops together and suspend each loop from the wood dowel or the glass rod on each beaker. Be sure the loop is immersed in the water, as in the diagram. After 20 minutes, test 5 mL of the water in each beaker for the presence of starch and 5 mL of water for the presence of simple sugar.

Complete this Investigation by answering questions 10–12.

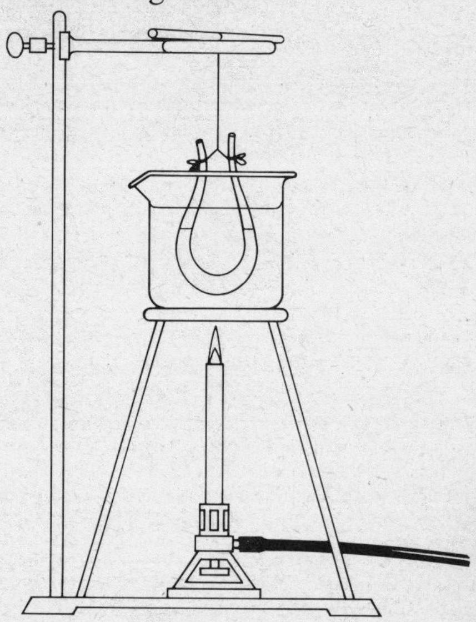

Investigations On Your Own

Gastric digestion involves the enzyme pepsin and chemically changes proteins. The gastric glands also manufacture hydrochloric acid. Make a solution of artificial gastric fluid by mixing 2 g of commercial pepsin in 0.4% hydrochloric acid to make 100 mL of the fluid. Filter this and test with litmus paper. Number 6 test tubes and put them in a rack. Put a few pieces of cracker in tube *1*, a small piece of hard-cooked egg white in tube *2*, and a small piece of butter in tube *3*. Fill each tube one-third full with the artificial gastric fluid that you have made. In tubes *4*, *5*, and *6*, put a piece of the same foods, but fill each tube one-third full of water.

Allow the 6 test tubes to stand overnight in a warm place. Check the cloudiness of each solution and explain your results.

NAME ______________________ DATE ______________________

41 Starch Digestion

DATA SHEET

Analysis

Trace and cut out 3 glucose models and 1 amylase model as directed in the Procedure page. Remember that the model is *not* an actual representation of the molecule.

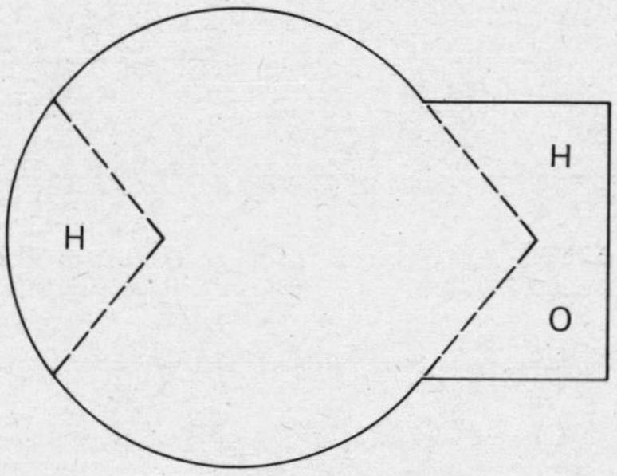

Glucose model (sugar part of starch molecule)

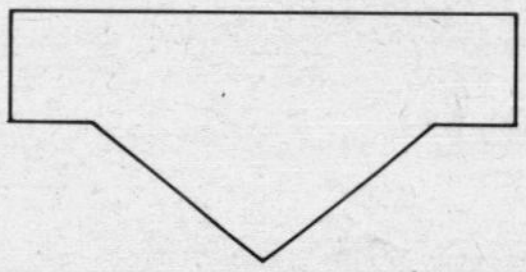

Amylase model (enzyme)

(1) By attaching glucose models together, you are forming a model of which larger molecule? ______________________

(2) What is lost when glucose models are joined together? ______________________

(3) What happens when amylase is fitted between the last two glucose molecules?

____

____

(4) If amylase is placed between the last remaining glucose models, what substance remains? ______________________

Investigation

(5) What happens when iodine is added to saliva? ______________________

(6) What happens when iodine is added to cooked starch? ______________________

____

(7) What do your results indicate? ________________

(8) List your results in the 3 test tubes. ________________

(9) Explain your results. ________________

(10) What substance passed through the membrane? ________________

(11) What substance did not pass through the membrane? ________________

(12) Describe the test in the beaker where saliva plus cooked starch solution was placed in the membrane. ________________

Summary

(1) What chemical changes occur when starchy foods are chewed? ________________

(2) Why is it important for the long chain of starch to be changed to the simple sugar glucose? ________________

Fill in the following blanks:

(3) Starch is a molecule that can be broken down by enzymes to form ________.

(4) Digestion involves ________________ large molecules.

(5) Saliva contains the enzyme ________________.

(6) ________________ C was used for some of these investigations because it is the temperature of the body.

(7) Simple sugar molecules are small enough to pass through ________________.

(8) ________________ is used in testing for the presence of simple sugars.

NAME DATE

Blood Types and Circulation 42

MATERIALS

INVESTIGATION▶

- human whole blood
- sterile lancets
- sterile cotton
- alcohol
- toothpicks
- blood-typing sera:
  anti-A
  anti-B
- wax marking pencil
- microscope slides
- microscope
- paper toweling
- Petri dish
- live goldfish
- medicine dropper

OBJECTIVES▶

- To understand the basis for blood typing
- To learn the technique for blood typing
- To observe blood circulating within a living system

Problem

How are blood types determined? How does blood circulate within a living system? Blood types (A, B, AB, or O) are inherited traits. Blood samples can be typed by using serum from type A and type B blood. If agglutination or clumping occurs when the test serum is added to the blood sample, the blood type can then be identified.

In humans and in many animals, blood moves in a system of closed tubes powered by the heart (pump). Arteries and arterioles carry the blood *away from* the heart; veins and venules carry the blood *toward* the heart. Thin-walled capillaries connect the arterioles and the venules.

Procedure

Part I. Analysis

Read the following paragraphs carefully. Compare the statements to the results in the Chart on page 259.

Blood types are inherited. Agglutination comes about because of a reaction between a protein in the serum and a protein on the red blood cell membrane. The protein on the red blood cell is called an *agglutinogen*. It is either A or B. Membranes or red blood cells of a person with type A blood contain A agglutinogen. Those of a person with type B blood contain B agglutinogen. Membranes of red blood cells from AB blood contain both A and B agglutinogen. The red blood cell membranes from type O blood do not contain either of these agglutinogens.

The protein in serum is called an *agglutinin*. Agglutinins are antibodies and have been given names to indicate the types of red blood cell protein that they cause to clump. A person with type A blood, containing A red blood cell antigen, would have anti-B serum antibody. This anti-B serum, then, would cause a clumping of red blood cells that contained the B antibody. A person with type B blood, containing B red blood cell antigen, would have anti-A serum antibody. Type AB blood has neither serum antibody present. Type O blood has both anti-A and anti-B serum antibodies present. Answer Data Sheet questions 1–5.

Part II. Investigation

(A) Now you will determine your own blood type. Use a wax pencil to divide a microscope slide in half. Label the left side "anti-A" and the right side "anti-B."

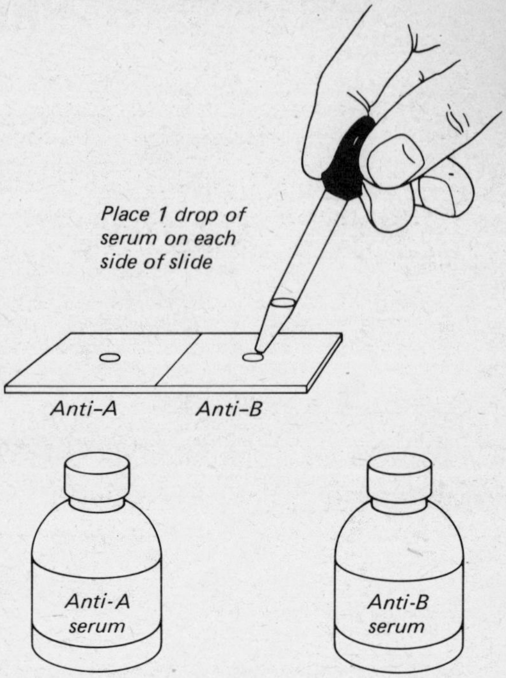

Put one drop of anti-A typing serum on the left side of your slide. Put one drop of anti-B typing serum on the right side of your slide.

Clean the tip of your middle finger with a swab of alcohol. Using a sterile lancet, pierce the skin at the tip of the finger. If blood does not flow at once, stroke the finger toward the tip, pressing as you stroke. (REMEMBER: *Do not use the same lancet for more than one student or more than once.*) Put one drop of blood from your finger (or provided by your teacher) on each side of the glass slide near the drop of typing serum. Do not dip your finger into the typing serum. Work quickly so the blood will not clot on the slide.

Use a toothpick to mix one drop of blood with the anti-A typing serum. *Use a different toothpick for mixing the other drop of blood with the anti-B typing serum*. After one minute, examine each side for clumping of cells. This clumping is called *agglutination*. Clumping will appear as cells cluster together. Unclumped blood will keep a smooth, even appearance. Compare your results with the Chart on page 259. Record your findings by answering questions 6–10.

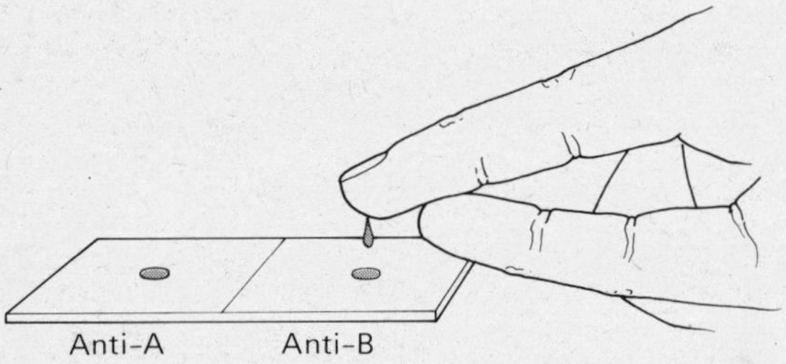

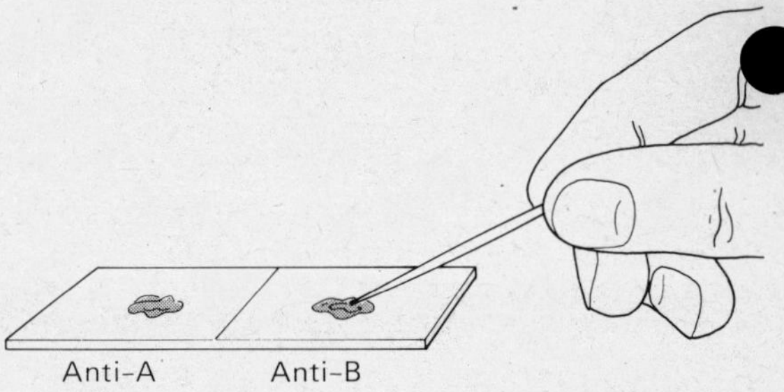

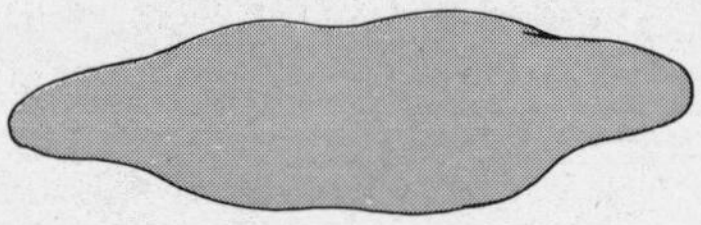

No agglutination

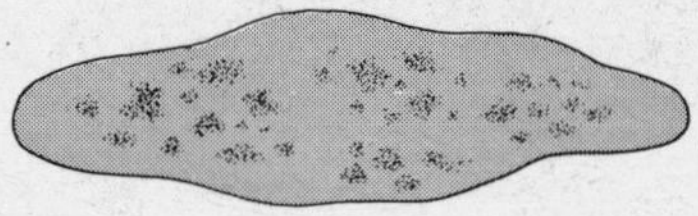

Agglutination (clumping)

(B) It is important to understand not only the components and typing of blood, but also the way in which blood circulates within a living system.

Saturate paper toweling or cotton with water. Gently lay a goldfish on a glass slide. Wrap the wet toweling around the middle of the fish and the slide. Leave the head of the fish and the tail fin uncovered.

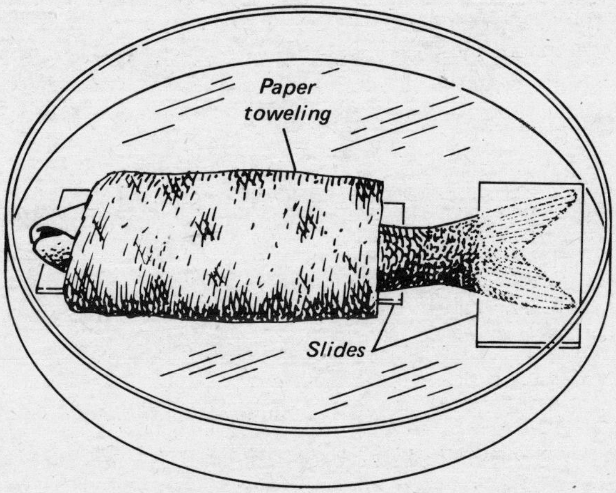

Place the wrapped fish in a Petri dish. Spread the tail fin and place a slide over it to hold it down. Use a dropper to keep the fish's skin moist during the Investigation.

Now place the Petri dish on your microscope stage. You may have to remove the stage clips to do this. Under low power, focus on the tail membrane. Observe the blood vessels in the tail.

If possible, observe the blood vessels under high power. Locate a blood vessel in which the cells are moving in single file. (This is a capillary.)

Record your findings by answering questions 11–15 on your Data Sheet.

Investigations On Your Own

(1) Examine prepared slides of blood smears. If prepared slides are not available, use the following procedure to prepare your own. Clean the tip of your middle finger with a swab of alcohol. Use a sterile lancet to pierce the skin and obtain a large drop of blood. [See safety instructions in Investigation, Part **(A)**.] Put a drop of blood on the slide. Use the edge of another slide to spread the blood into a thin smear. Allow the smear to dry for a few minutes, then pass it quickly through a flame 3 times. (This will warm the slide and fix the blood cells in place.) Add a few drops of Wright's stain to the smear. After 2 or 3 minutes, rinse the stain from the slide with water. Dry the slide. Now study the slide under the high power of your microscope. Look at the different kinds of blood cells and the spaces between them. Refer to your text to help you identify the different components of blood.

(2) Observe red blood cells in various saline solutions to determine the salt concentration within the cells. In doing this activity, you may also determine what may happen to cells in the following solutions: 0.3%, 0.85%, and 1.5%.

(3) Make a study of the factors involved in the clotting time of blood. Prepare and give a demonstration to the class on the procedure of measuring clotting time. Explain factors that can cause changes in the clotting time.

(4) Make a large chart showing how the various blood types are inherited. Although your blood type is determined by two genes, there are three possible genes or alleles you may inherit. They include I^A for type A blood, I^B for type B blood, and I^O for type O blood. I^A and I^B lack dominance with one another, so a person with both of these genes has AB blood. I^O is recessive to both I^A and I^B, so a person having I^AI^O has type A blood and one having I^BI^O has type B blood. Type O can only occur when the two genes for O are present.

If:	*Anti-A*	*Anti B*	*Blood type*
Agglutination in anti-A only	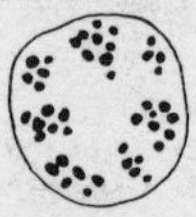	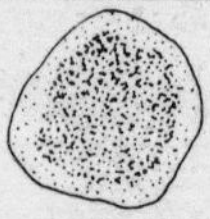	A
Agglutination in anti-B only	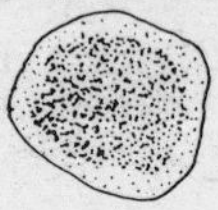	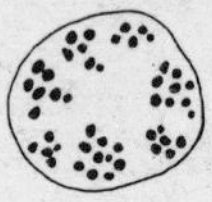	B
Agglutination in both	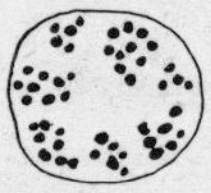	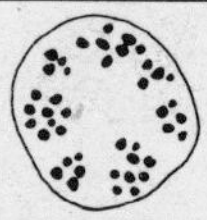	AB
No agglutination	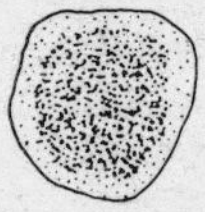	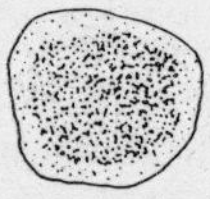	O

NAME DATE

Blood Types and Circulation 42

DATA SHEET

Analysis

Refer to the information on your Procedure page and the Chart on the opposite page in order to answer the following questions.

(1) With which blood types would the anti-B serum cause clumping? ______

(2) With which blood types would the anti-A serum cause clumping? ______

(3) What would happen if either anti-A or anti-B serum protein were added to type AB blood? ______

(4) What would be the result in blood typing if anti-A typing serum were added to type O blood? ______

(5) What would be the result if anti-B were added to O blood? ______

Investigation

(A) Determine your blood type by following the instructions on your Procedure page.

(6) What were your results on the anti-A side? ______

(7) What were your results on the anti-B side? ______

(8) What is your blood type (based on your findings)? ______

(9) Do you have any agglutinogens (antigens) on your red blood cell membranes? Explain your answer. ______

(10) Do you have any agglutinins (antibodies) in your serum? Explain your answer. ______

(B) Study how the blood circulates by observing the movement of blood in a live goldfish.

(11) Why is it important to keep the fish's skin moist during the Investigation? ______

______

(12) Describe what you see under the low power of the microscope as you focus on the fish's tail. ______

(13) Describe the flow of blood. ______

(14) What is the function of the capillaries? ______

______

(15) How do capillaries differ from other blood vessels? ______

______

Summary

Fill in the following chart.

Blood Type	Agglutinogen Present on Red Blood Cell	Agglutinin Present in Serum
A		
B		
AB		
O		

Write the correct blood type(s) before each of the following statements.

______ **(1)** anti-A and anti-B serum antibody present

______ **(2)** A agglutinogen on red blood cells

______ **(3)** no agglutinogen on red blood cells

______ **(4)** cells agglutinate in anti-A typing serum

______ **(5)** cells agglutinate in anti-B and anti-A typing sera

______ **(6)** no agglutination of cells in either anti-A or anti-B typing sera

______ **(7)** cells agglutinate in anti-B typing serum

______ **(8)** A and B agglutinogen on red blood cells

______ **(9)** anti-B serum protein present

______ **(10)** B agglutinogen in red blood cells

NAME DATE

43 Lung Capacity and Carbon Dioxide Production

MATERIALS

ANALYSIS▶
- watch or clock with second hand

INVESTIGATION▶
- round balloon
- metric tape measure
- phenolphthalein solution
- 125 mL Erlenmeyer flask
- dropper
- 6 mm diameter straws
- 0.04%NaOH solution in a labeled dropper bottle
- 0.4% NaOH solution in a labeled dropper bottle

OBJECTIVES▶
- To measure lung air capacity
- To determine relative amounts of CO_2 in exhaled air
- To compare CO_2 production at rest and after activity

Problem

How is lung capacity measured? How can the relative amount of carbon dioxide in exhaled air be determined?

An average lung capacity is about 4,500 milliliters or 4.5 liters. During normal breathing, 500 milliliters of air are inhaled and exhaled. Forced breathing increases the amount of air moved in and out of the lungs.

Carbon dioxide is a waste product of breathing. The amount of carbon dioxide released increases as activity (energy demand) increases. By using phenolphthalein in a sodium hydroxide solution, the presence of carbon dioxide can be determined.

Procedure

Part I. Analysis

Using your textbook or other reference book, review the structures through which air passes during breathing. While sitting quietly, measure (or have someone else measure) your respiration rate for one minute. The respiration rate is the movement of your ribs, up and down, in your chest. Note what happens during inspiration and expiration.

The air moving in with each inspiration and out with each expiration is called *tidal air*. Observe your own breathing. At the end of normal inspiration, attempt to continue taking in air instead of expiring. Take note of your results.

The volume of air that can be inhaled above a normal inspiration is called *complemental air*. Force your intake of air and then breathe normally again. Note how much air you can now exhale. This is called *reserve air*.

Vital capacity is the sum of tidal, complemental, and reserve air. Some air remains in your lungs and cannot be forced out under normal conditions. This is called *residual air*. Note the Table on your Data Sheet that shows the *average* volumes of lung capacity. Record your observations of respiration rate and lung capacity by answering Data Sheet questions 1–6.

Part II. Investigation

(A) In Analysis, you observed the effect of respiration rate on your lung capacity. Now you will measure your lung capacity under various conditions.

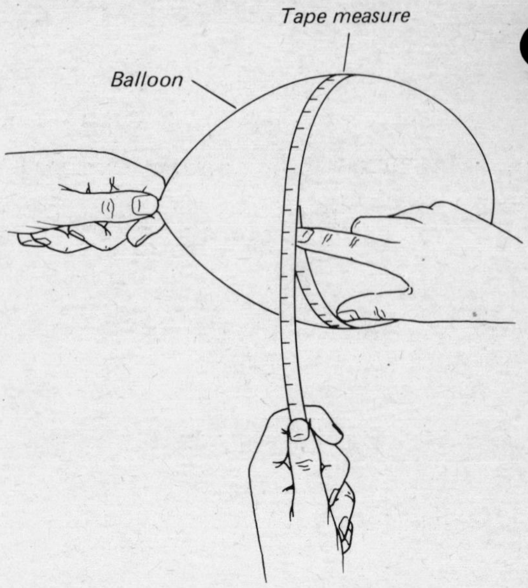

Prepare a round balloon for this Investigation by stretching it a few times. Then blow it up once and let the air out.

Try to blow up the balloon with the amount of air that you normally exhale. Carefully pinch the balloon to prevent air from escaping. Now measure the circumference of the balloon, using a metric tape measure. This amount of air is approximately equal to the tidal capacity of your lungs. Record this tidal capacity number in the Data Sheet Table. Deflate the balloon and then take two more readings to find an average measurement.

Now breathe normally. Then force as much air as possible into the balloon. Measure and record the circumference of the balloon in the Data Sheet Table under the heading of reserve air volume. Deflate the balloon. Take two more measurements to find your average reserve air capacity.

Finally, take a deep breath and blow up the balloon with as much air as you can exhale. Measure and record this vital capacity circumference in the Data Sheet Table. Repeat your measurement of vital capacity two more times. Calculate your average vital capacity.

Compare your measurements with those of your class. Record the (class) average tidal capacity, reserve air volume, and vital capacity. Then answer Data Sheet questions 7–8 based on your findings and on the information in the Analysis section.

(B) Carbon dioxide is a waste product of respiration. By performing the following experiment, you will be able to determine if the rate of CO_2 production changes as activity increases. Refer to the drawing on the next page as you read the procedure instructions.

Put 100 mL of water in an Erlenmeyer flask and add 5 drops of phenolphthalein. Add 0.04% NaOH solution drop by drop until a faint pink color remains in the solution in the flask. Put a straw into the flask and breathe *in* through your nose and *out* through the straw. Be careful to avoid sucking any solution into your mouth. The exhaled breath will bubble up through the solution. Continue this procedure (with your normal breathing rate) for one minute. Note any changes in the color of the solution.

CAUTION: *NaOH can cause a burn. Always place the dropper back in the bottle.*

NAME DATE

Now remove the straw and add 0.4% NaOH one drop at a time. Swirl the contents of the flask after each drop. Stop when the solution remains pink for one minute.

Record the number of drops of 0.4% NaOH in the Data Sheet Table. (This test not only indicates the presence of CO_2, it indicates the amount. The more drops of NaOH required, the more CO_2 is present.)

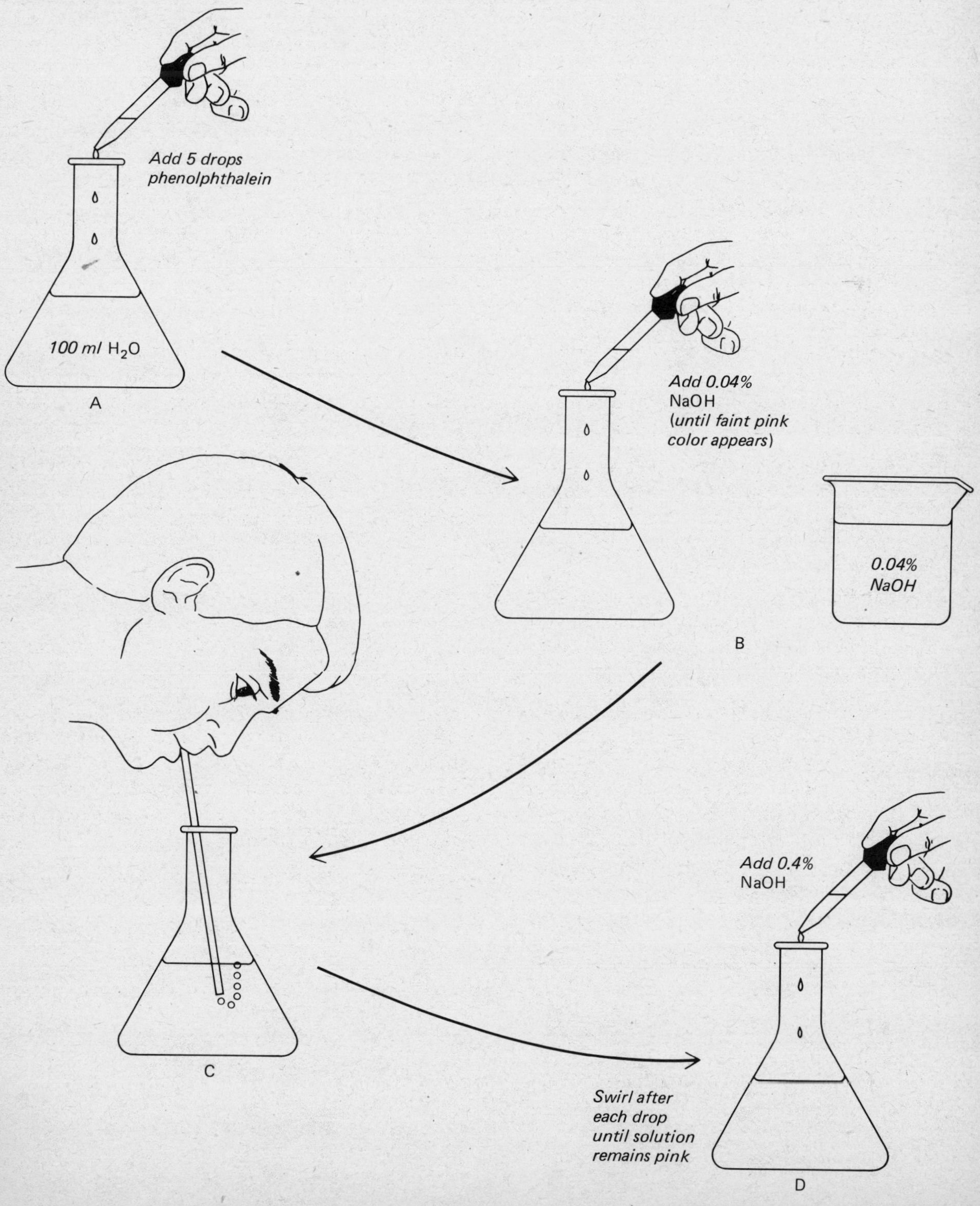

Rinse out the flask and prepare it for another test. Add 100 mL of water to the flask. Then add 5 drops of phenolphthalein and 0.04% NaOH drop by drop until a faint pink color remains.

Run in place for one minute. Now, using a straw again, exhale into the flask for one minute. While you are doing this exhaling, have your laboratory partner count your respiration rate. Compare your respiration rate after exercise with your respiration rate before exercise.

Add the 0.4% NaOH solution until the flask remains faintly pink for one minute. Count the number of drops required and record this in the Data Sheet Table.

With the help of your teacher, find out the average class measurement of CO_2 (based on the number of drops of NaOH needed to neutralize) before and after exercise. Complete this Investigation by answering Data Sheet questions 9–10.

Investigations On Your Own

(1) Use the same procedure as in Investigation **(B)** to determine carbon dioxide production under various activities. Try to predict the effect of a vigorous gymnastic class, a track meet (as a participant or a spectator), or a public speaking situation. Test your predictions under varying conditions. Then reach an educated conclusion about the production of CO_2.

(2) Determine whether any differences in carbon dioxide production can be related to differences in physical condition. Divide a group of students into those who exercise regularly and those who do not exercise regularly. Do you find variations in carbon dioxide production in these two groups?

NAME DATE

Lung Capacity and Carbon Dioxide Production 43

DATA SHEET

Analysis

Use your text and your observations to answer the following questions about air movement and lung capacity.

(1) How many times do you inhale in one minute? ______

(2) Describe what happens during inspiration. ______

(3) Describe what happens during expiration. ______

(4) What happens when you try to take in more air rather than exhaling? ______

(5) Is it possible to exhale more air than you take in? Explain. ______

Note the figures for lung capacity in the table.

(6) What individual differences might account for *normal volumes* differing from the averages given in the Table?

LUNG CAPACITY	
tidal air	0.5 L
complemental air	2.5 L
reserve air	1.5 L
vital capacity	4.5 L

Investigation

(A) Measure your lung capacity under varying conditions by doing the balloon activity described in your Procedure page. Record your results in the Table on the next page.

Lung Capacities (Balloon circumference in cm)

Trial Reading	Tidal Capacity	Reserve Air Volume	Vital Capacity
1			
2			
3			
Average			
Class Average			

(7) How can you account for the difference in tidal capacity and reserve air volume?

____

(8) How do your capacities compare with the class average capacities? ______________

____

(B) Fill in the following Table based on the number of drops of NaOH needed to neutralize the solution.

NUMBER OF DROPS OF 0.4% NaOH NEEDED TO NEUTRALIZE		
Activity	**Your Measurement**	**Average Measurement for the Class**
Resting		
Running in place		

(9) Compare your resting respiration rate in Analysis to the respiration rate after exercise. ____

(10) What does the change in respiration rate indicate about the *volume* of air exchanged before and after exercise? ______________________________

Summary

(1) What is vital capacity? Differentiate between tidal air, complemental air, and reserve air. ____

____

____

____

(2) What is the relationship between CO_2 production and respiration rate? ______________

____

NAME DATE

Vision and Other Senses 44

MATERIALS

ANALYSIS▶
- penlight flashlight

INVESTIGATION▶
- index card
- colored pencils or crayons
- white unlined paper
- 10% sugar solution
- 10% vinegar solution
- 3% salt (NaCl) solution
- 0.001% quinine sulfate solution
- cotton swabs
- vanilla extract
- apple
- onion

OBJECTIVES▶
- To identify the major parts of the eye
- To observe and identify particular characteristics of vision
- To determine the location of taste receptors
- To understand the role of smell receptors

Problem

How do we see? How can we test the senses of taste and smell?

Our vision depends upon light-sensitive cells of the retina and the way in which the brain interprets what the eye sees. Each of our senses — seeing, hearing, tasting, smelling, and touching — involves a stimulus and a response.

Taste and smell are two senses that work together. The cells on the tongue and the roof of the mouth can detect sweet, salty, bitter, and sour. However, when the sense of smell is blocked, our taste buds have difficulty distinguishing flavors.

Procedure

Part I. Analysis

Examine the drawing of the eye on your Data Sheet page. Use charts of the eye and/or your textbook to label the following structures: *pupil, lens, iris, cornea, retina, optic nerve, fovea*.

Have someone look closely at your eye and observe the size of your pupil. Next, ask your partner to shine a flashlight into your right eye while looking at the pupil. Note what happens to the size of the pupil. Now close your eyes and cover them lightly with your hands for 30 seconds. Note what happens to the pupils of your eyes as you open and uncover your eyes.

If possible, repeat this test, with your partner as the one tested and yourself as the observer. Do your results agree? Answer Data Sheet questions 1–3.

Part II. Investigation

(A) Look closely at the drawing of the eye in the Analysis section of the Data Sheet. Note that the projection of the pin is inverted on the retina. However, when the receptors of the retina stimulate nerve cells and the impulse is carried to the brain, we interpret the image in its correct position.

Do the following activity to demonstrate that the image is actually reversed. Punch a small hole in an index card with a sharp pencil. Hold the card up toward a light source. Place the card about 10 cm from your eye and look through the hole with your left eye, closing your right eye. Carefully hold the pencil point close to your left eye in the beam of light from the hole.

Move the pencil tip to the right and note the way in which the pencil appears to move. Move the pencil tip slowly up and down. Note the image that you see in the hole.

Answer Data Sheet questions 4–7.

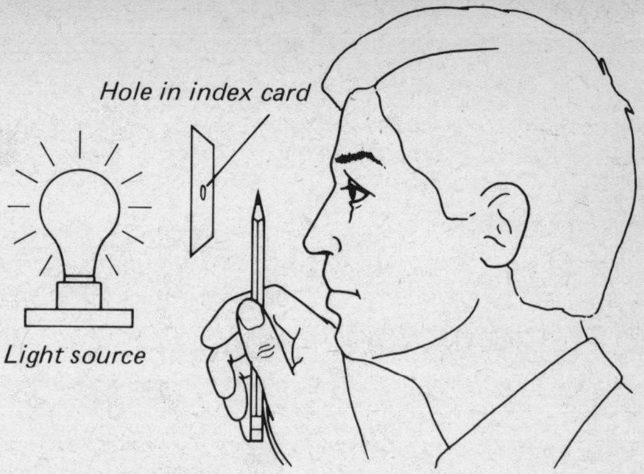

(B) Look at the two drawings of butterflies on your Data Sheet. Use colored pencils to color the labeled sections on the left figure. Now stare at the dot in the center of the butterfly with the colored wings. After 30 seconds, stare at an unlined sheet of white paper. Observe the colors that appear. You may have to do this several times to identify the colors. Now, in the right-hand butterfly figure, color the sections with the colors you perceived in the *afterimage* on the white paper. The phenomena you've just experienced is due to retinal fatigue. Answer Data Sheet questions 8–9.

(C) In this Investigation, you will find the location of taste receptors. Look at the diagram of the tongue and identify the 4 areas to be tested.

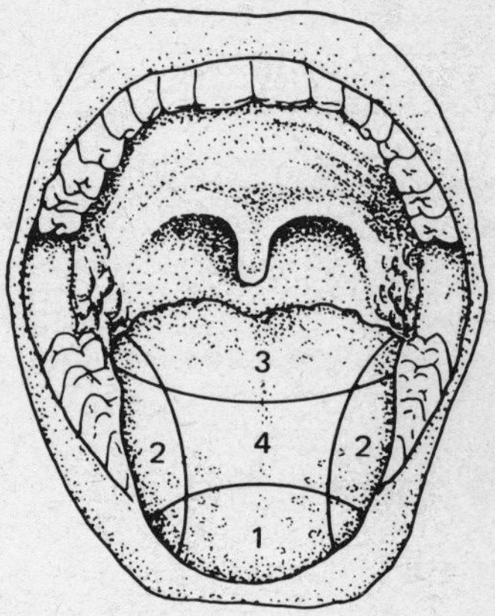

Rinse out your mouth with water and pat your tongue dry with a paper towel. Your lab partner will dip a clean cotton swab into the 10% sugar solution. Extend your tongue, and your lab partner will touch each area (one at a time) with the sugar solution. Tell your partner when you perceive the "sweet" taste.

CAUTION: *Dip the cotton swab into the sugar solution only once. Use the same swab to test each area of the tongue.* When the areas have been tested, discard the swab to prevent comtamination. Follow the above procedure for the vinegar solution, the salt solution, and the quinine sulfate solution. Record your findings in the Data Sheet chart. Repeat your testing, using your partner's tongue, and record the results.

Now hold your nose closed with your thumb and forefinger. Have your partner test the areas of your tongue with a swab that has been dipped into vanilla extract. Note the taste you perceive and the area of your tongue where you sense it. Compare your sense of taste with your sense of smell by smelling the vanilla. Answer question 10.

Finally, try to distinguish certain foods that you cannot smell. Use samples of onion and apple that have been cut into small pieces. Blindfold some volunteers and have them hold their noses. Let them taste one of the foods. After they have swallowed the first sample, have them taste the other. Answer question 11.

Investigations On Your Own

Devise an experiment to test the sense of touch to various temperatures of water. Set up at least three containers, each with either ice cold water, lukewarm water, or hot water. Try to find which parts of the hand are most sensitive to cold or heat. Note also the effect of immersing your hand in the hot and then in the lukewarm water. Is the sense of touch relative?

NAME DATE

44 Vision and Other Senses

DATA SHEET

Analysis

Label the parts of the eye as directed on your Procedure page.

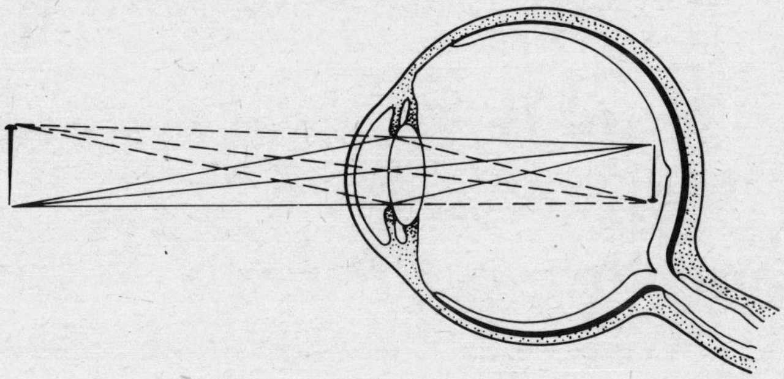

(1) What happens to the pupil of the eye as a light is shined in your eye? ______

______

(2) Describe the effect of closing your eyes for 30 seconds and then opening them.

______

(3) What do you think is the function of the pupil? ______

______

Investigation

(A) Perform the activity described on your Procedure page to find out if an image is actually reversed.

(4) When you move the pencil point to the right, which way does it appear to move?

______

(5) Describe the appearance of the pointed pencil. ______

______

(6) How would you explain your answers to **(1)** and **(2)**? ______

______

______

______

(7) In order to see the pencil point right-side up, what would you have to change in this activity? ____________________

(B) Color the labeled sections of the butterfly figure on the left. Then fill in the butterfly on the right with the colors you perceived in the afterimage.

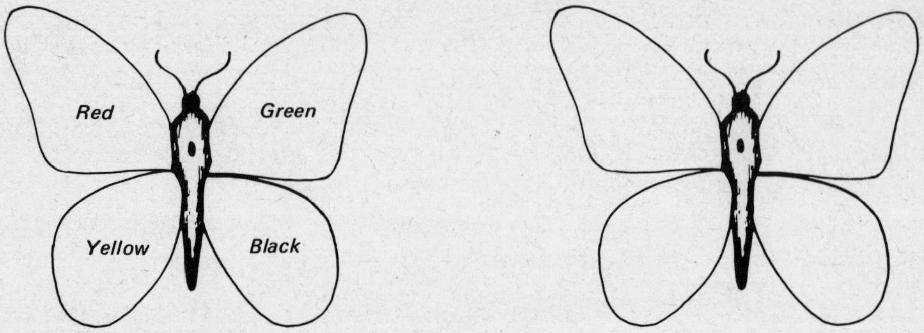

(8) Describe the colors observed in the afterimage. ____________________

____

(9) Why is retinal fatigue a good description for this activity? ____________________

____

____

(C) Complete the Table below, indicating the area of the tongue sensitive to each taste sensation.

AREA OF THE TONGUE		
Substance	**You**	**Partner**
sugar		
vinegar		
salt		
quinine sulfate		

(10) What was unusual about the taste and smell of vanilla extract? ____________________

____

(11) Was it possible to distinguish the onion and apple? ____________________

____

Summary

(1) In all of our senses, there must be a <u>stimulus</u> received by a <u>receptor</u> and a <u>response</u> to the stimulus. Explain what the underlined words mean. ____________________

____

____

NAME DATE

Alcohol and Tobacco 45

MATERIALS

INVESTIGATION▶

- prepared slide of normal lung tissue
- prepared slide of cancerous lung tissue
- microscope
- cigarettes, variety of filtered and unfiltered
- one-hole rubber stopper
- two-hole rubber stopper
- Buchner funnel
- Erlenmeyer flask with side vent
- rubber tubing
- 2 right-angle glass tubing pieces
- large jug
- large pan or pail

OBJECTIVES▶

- To analyze statistical data concerning the abuse of alcohol
- To examine normal and cancerous lung tissue samples
- To simulate the effect of smoking on lungs by means of a smoking machine and filter paper

Problem

How does the use of alcohol affect the human body? What are the physical effects of smoking tobacco?

Alcohol is a drug that acts as a depressant on the nervous system of the body. As alcoholic content increases, there is a loss of judgment, relaxing of tensions, blurred vision, slurred speech, dehydration of tissues, difficulty in muscular coordination, etc. The abuse of alcohol — alcoholism — is a disease.

Smoking causes tissue damage in the lungs that may lead to respiratory disease such as bronchitis, emphysema, and lung cancer. Smoking has also been linked to circulatory problems, heart disease, and peptic ulcers.

Procedure

Part I. Analysis

Read the following United States' statistical data on alcoholism. Then answer Data Sheet questions 1–3.

. . . 3.3 million young people (ages 14–18) are problem drinkers, alcoholics, or potential alcoholics.
. . . 10 million people are alcoholics or problem drinkers; there were only 5 million ten years ago.
. . . Alcohol is the number one cause of accidental deaths in the age group 15–24.
. . . Almost 50% of all highway fatalities have alcohol as an underlying cause.
. . . 60% of the people killed in drunken driving accidents are teenagers.
. . . Alcoholics shorten their lives by 10–15 years.
. . . 55% of all arrests are linked to the abuse of alcohol.

Part II. Investigation

(A) Examine a prepared slide of normal lung tissue under the low-power objective of your microscope. Then switch to the high-power objective. Draw what you observe in the space provided in your Data Sheet.

Now examine a prepared slide of cancerous lung tissue. Note the differences between this slide and the normal lung tissue sample. Focus under low power, then switch to the high-power objective. Draw what you see in the space provided on your Data Sheet. Label each drawing. Answer questions 4–6 based on your microscopic observations and information contained in the Data Sheet bar graph.

(B) Set up a smoking machine apparatus as shown below. Test various kinds of cigarettes. Examine the filter paper after each cigarette is "smoked." Then compare the residues left on the filter papers. Answer Data Sheet questions 7–11.

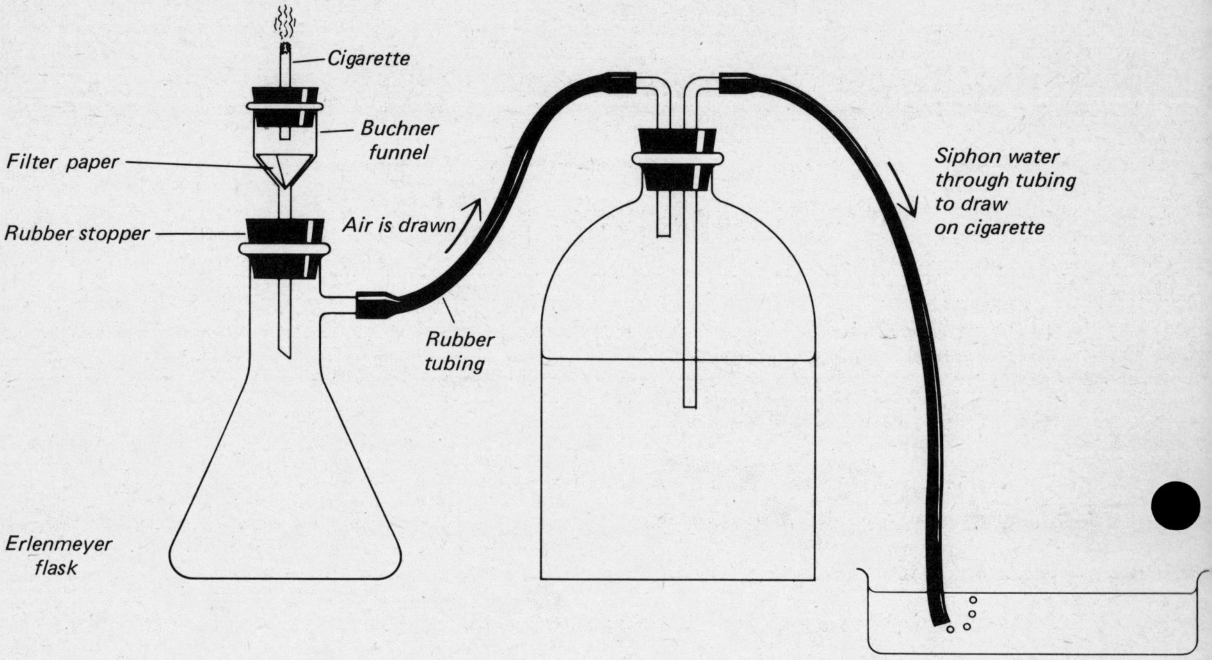

Investigations On Your Own

(1) Do a research report on the various kinds of drugs (stimulants, depressants, hallucinogens, etc.) involved in drug abuse. Identify their chemical and common names, and their effects on the human body. Cite reasons why people turn to drugs. Include a report on programs designed to help drug addicts recover.

(2) Repeat the pulse and breathing rate activities from Lab 40 with a group of smokers and nonsmokers. Evaluate your results. Do your findings agree with scientific data indicating the harmfulness of smoking?

NAME DATE

Alcohol and Tobacco 45

DATA SHEET

Analysis

(1) List several reasons why people drink alcohol. ______________________

____

____

(2) What harm can alcohol do to people who abuse it? ______________________

____

____

____

(3) Which statistic on your Procedure page helped you to realize the seriousness of alcoholism? Explain your answer. ______________________

____

____

Investigation

(A) Examine the lung tissue slides and draw what you observed in the spaces below.

Incidence of lung cancer

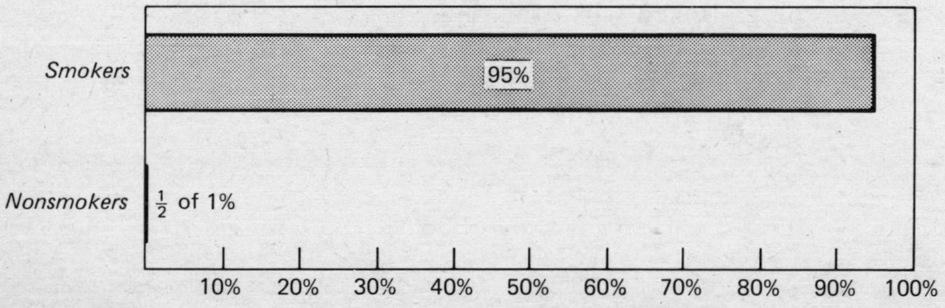

(4) What is the most significant difference between the normal lung tissue and the cancerous lung tissue? ______

______

______

(5) What is the purpose of cilia in lung tissue? ______

______

(6) Compare your microscopic observations with the statistical evidence in the bar graph. What do you conclude? ______

______

(B) Perform the smoking machine activity with various kinds of cigarettes as described in your Procedure page.

(7) Describe the residue collected on the filter papers. ______

______

(8) Which cigarette left the least amount of residue on the filter paper? ______

______

(9) Which cigarette left the most amount of residue? ______

(10) The filter paper corresponds to what part of a smoker's body? ______

______

(11) What do you conclude about the effects of smoking? ______

______

Summary

(1) Why is alcohol classified as a drug? ______

______

(2) Describe the effects of alcohol on the body. ______

______

(3) What diseases have been linked with smoking? ______

______

(4) What are some of the warning signs of alcoholism? ______

______

______

NAME DATE

The Effect of Hormones 46

MATERIALS

INVESTIGATION▶

- textbook
- 25 tadpoles with hind limb buds visible *(Rana pipiens)*
- 5 fingerbowls (9-cm diameter)
- pond water
- 5 Petri dishes
- 3 sheets of millimeter graph paper
- Thyroxin Stock Solution
- Iodine Stock Solution
- 1 jar of strained baby food (plain spinach)
- 2 graduated pipettes (1-mL)
- 100-mL graduated cylinder
- labels
- medicine droppers

OBJECTIVES▶

- To understand that hormones help to regulate bodily processes
- To observe how thyroxin influences metamorphosis in tadpoles

Problem

How do hormones regulate bodily processes? How do hormones affect development? Hormones are chemicals secreted by endocrine glands. Hormones act as chemical messengers, communicating instructions between different parts of the body. Some hormones help to maintain a balance; others are vital to growth and development.

The thyroid gland is one of the endocrine glands essential for normal growth. The hormone produced by the thyroid is called thyroxin. For the thyroid gland to produce thyroxin, enough iodine must be available. Thyroxin also regulates an organism's metabolism — the rate at which food energy is used.

Procedure

Part I. Analysis

Read each statement on your Data Sheet and then fill in the name of the endocrine gland responsible for that particular effect. Use your textbook or other reference material to help you. Remember that difficulties are caused by an excess as well as a deficiency of a specific hormone.

Part II. Investigation

Biologists have found that the thyroid hormone affects metamorphosis in tadpoles. Set up the following experiment to see the effect of an excess of thyroxin on tadpole development. You will do this by adding thyroxin to the water in which the tadpoles are living.

The thyroid gland requires iodine to produce its hormone. If the iodine supply is too low, growth and metamorphosis in the tadpole is retarded. If we were to increase the supply of iodine in the water, would the tadpoles make more thyroid hormone and would metamorphosis be accelerated? Test the effects of added thyroxin and the effects of added iodine.

Number 5 fingerbowls and put 100 mL of pond water into each. Then put 5 tadpoles into each bowl. Change the water every day and treat the bowls as shown in the drawings below.

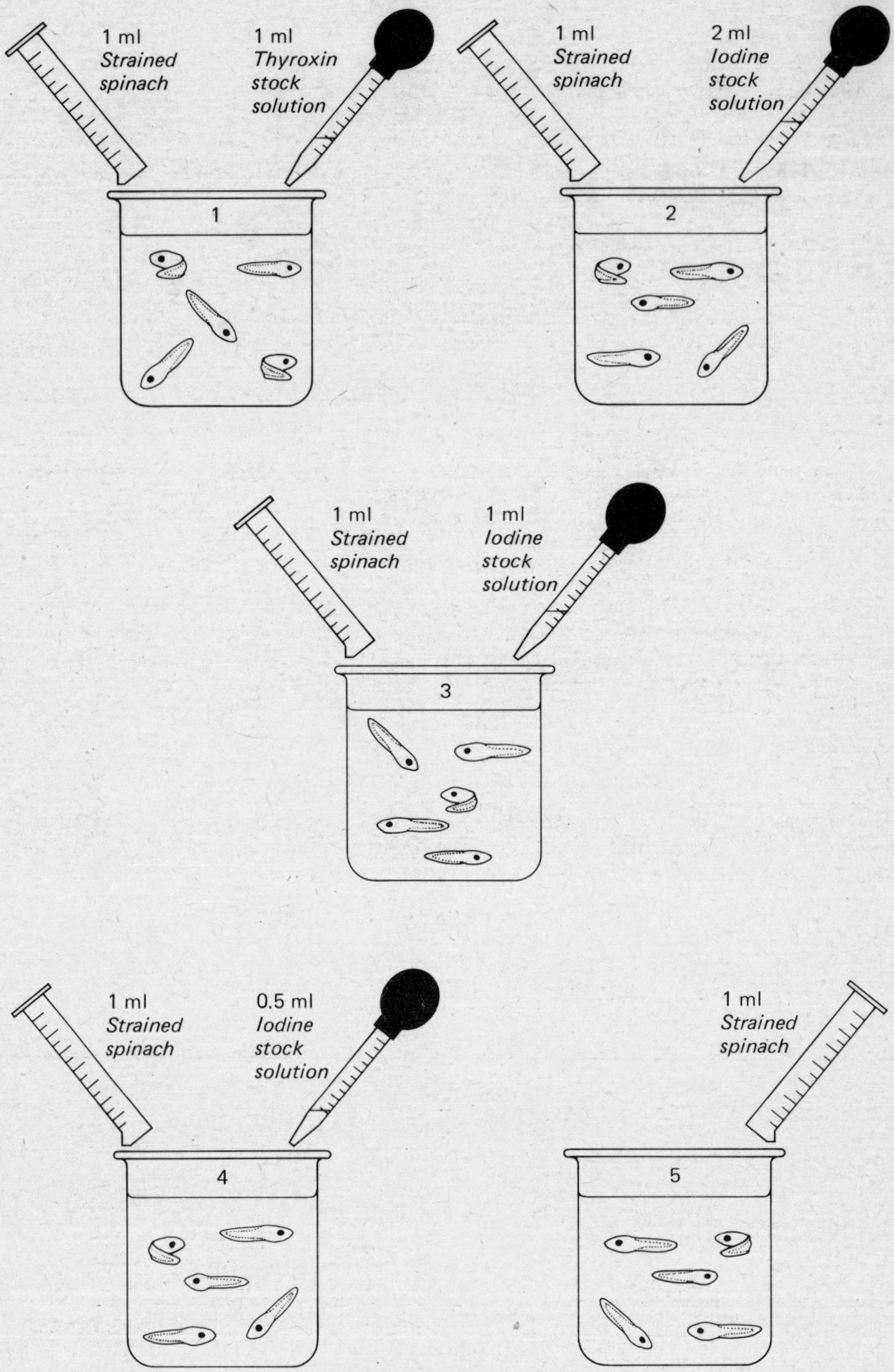

Maintain the cultures at room temperature.

Tape a piece of millimeter graph paper to the underside of each of 5 Petri dishes. Number these to correspond to the numbers of the fingerbowls. *Always use the same Petri dish for the same animals*.

Carefully remove the 5 tadpoles from fingerbowl 1 and place them in Petri dish 1. Measure and record total length, body length, and tail length. The tail length is the total length minus the body length. Return the animals to their fingerbowl immediately. Note the instructional drawing below. Repeat the measurements with fingerbowls 2, 3, 4, and 5.

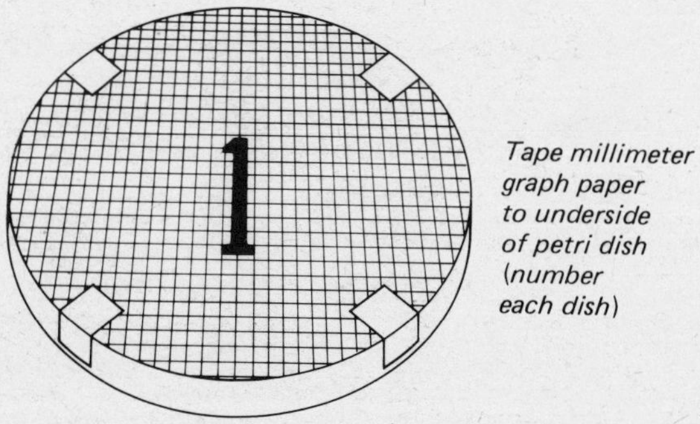

A

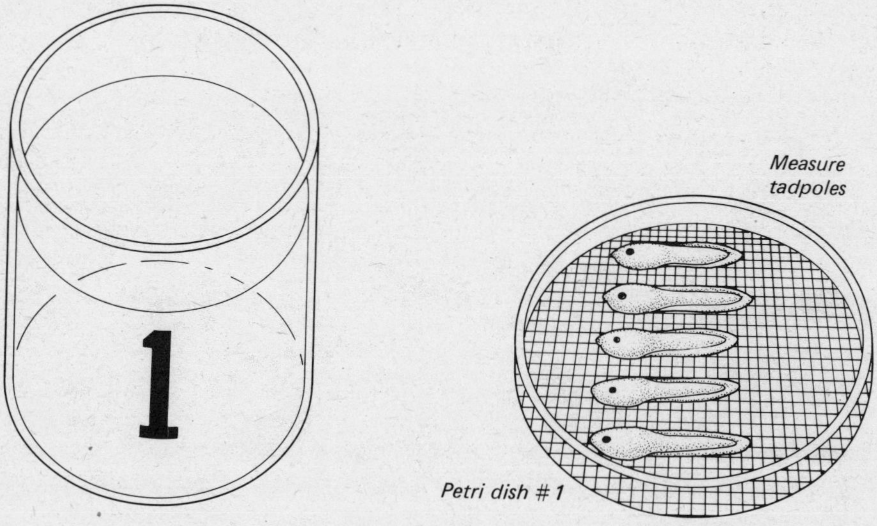

B

Although the medium in each of the bowls should be changed daily, these measurements can be made every 2 to 3 days. The total time for this Investigation will vary, but the average will be between 2 to 3 weeks. Be certain to observe the changes that indicate metamorphosis. In addition to body length, note the growth of hind- and forelimbs, the development of joints in the feet, changes in body shape, and changes in the size and shape of the mouth. Answer questions 11–12.

Record your measurements in the Data Sheet Table. Then make a graph of the mean body length of each group against the days observed.

Investigations On Your Own

Testosterone is a hormone produced by the testes that regulates the degree of maleness in an individual. Obtain at least two male chicks from a hatchery or egg farm. Rub some testosterone ointment on the comb area of the young chick. (The comb of a rooster is a sign of maleness.) Observe the comb growth of the "treated" chick and the untreated chick. Measure and record your results. Do you find a significant difference in comb growth?

NAME _______________ DATE _______________

The Effect of Hormones 46

DATA SHEET

Analysis

Identify the endocrine gland that applies to each statement. The endocrine glands include the following: *thyroid, parathyroids, pituitary, adrenal glands, pancreas, ovaries,* and *testes*. (You may use the name of a gland for more than one statement.)

_______________ **(1)** When angry or in danger, these glands secrete a hormone to give us extra strength.

_______________ **(2)** The islets of Langerhans on this gland secrete the hormone insulin.

_______________ **(3)** Female sex hormones estrogen and progesterone help to produce feminine traits.

_______________ **(4)** Too much growth hormone from this gland can result in a giant.

_______________ **(5)** Bone growth and muscle tone depend on the calcium balance controlled by this gland.

_______________ **(6)** Iodine deficiency sometimes causes an enlargement of this gland, which needs iodine for its hormone production.

_______________ **(7)** The small gland at the base of the brain that influences the activities of many other glands.

_______________ **(8)** Male traits — deep voice, beard, broad chest, etc. — are activated by the hormones from this gland.

_______________ **(9)** In one kind of diabetes, excess sugar is excreted in the urine. This is due to a lack of a hormone from this gland.

_______________ **(10)** The hormone secretions from this gland control the rate of metabolism — the use of food energy.

Investigation

Find the effect of varying concentrations of iodine and thyroxin solutions on tadpole metamorphosis by performing the experiment described on your Procedure page.

(11) Was there a significant difference in development between the tadpoles in plain water and those in iodine solution? _______________

____

(12) Describe some of the changes you observed in tadpole metamorphosis. _______________

____

____

____

Record the tadpole measurements in the Table below.

AVERAGE LENGTH MEASUREMENTS														
#1			#2			#3			#4			#5		
total	body	tail	total	body	tail	total	body	tail	total	body	tail	total	body	tail

Graph the body lengths of the tadpole groups according to the number of days from the beginning of the experiment.

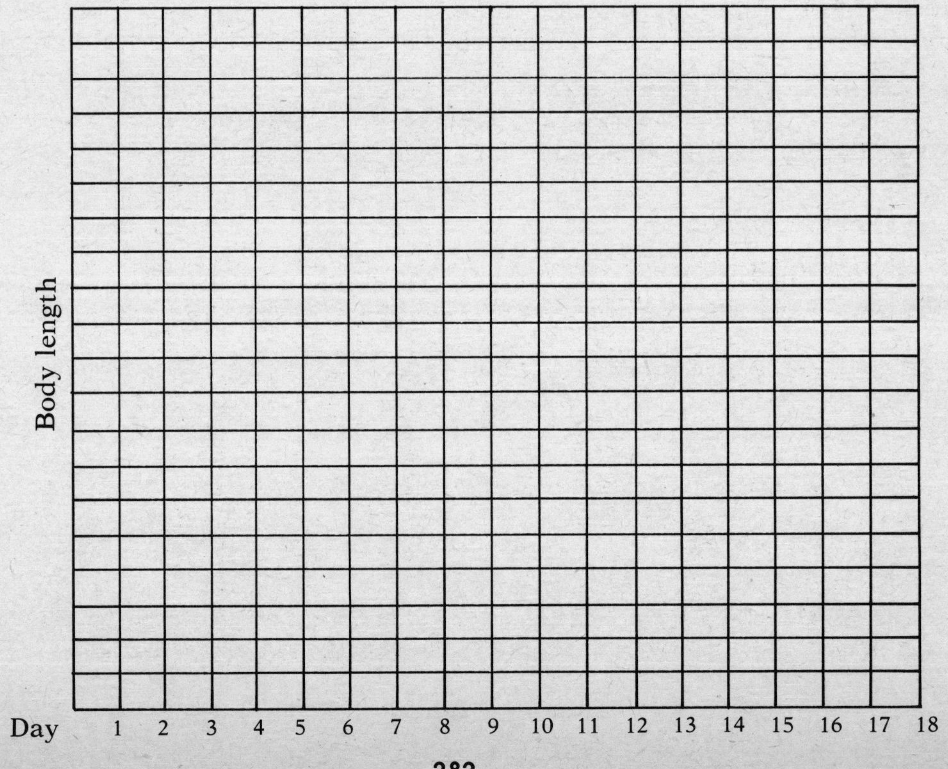

NAME DATE

Chick Embryo Development 47

MATERIALS

ANALYSIS▶
- unfertilized egg, bowl

INVESTIGATION▶
- fertilized egg (not incubated)
- fertilized egg (60 hours)
- fertilized egg (120 hours)
- cotton
- eye dropper
- scissors
- forceps
- fingerbowl
- dissecting microscope

OBJECTIVES▶
- To understand the structure of an unfertilized egg
- To observe a chicken egg at two stages of development
- To determine the function of the extraembryonic membranes
- To relate the developing chicken embryo to the human embryo

Problem

What is the structure of an egg? How does an embryo develop?
A chicken egg is made up of an egg cell (the yolk, nucleus, and cytoplasm) and the shell and shell membranes that protect against injury and loss of water.

After fertilization, the egg cell gradually develops into a young chick. A warm temperature — usually provided by the female parent or artificially by incubation — is necessary for development to take place. Embryonic development is a series of cell divisions, tissue and organ development, and growth until the young chick is ready to hatch. Chick development, from fertilized egg to hatching, takes 21 days.

Procedure

Part I. Analysis

Gently break an egg into a small bowl of water. Examine the yellow *yolk* and look for a small white spot on the yolk — the *blastodisc*. The clear fluid surrounding the yolk is the *albumen* or egg white. Note the cord of albumen on each end of the yolk. This is the *chalaza*.

In the space provided on your Data Sheet, draw the parts of the egg you have observed. Label the yolk, blastodisc, albumen, and chalaza.

Part II. Investigation

(A) Prepare a bed of cotton in a fingerbowl and place a fertilized, unincubated egg on the cotton so that the egg is held in position. Expose the egg by cutting the shell as shown in figures A and B on the following page. When the cutting is completed with an opening large enough to permit examination, carefully remove the cut shell with a forceps.

Observe the large yellow *yolk*, which contains stored food for the developing embryo. On the upper surface of the yolk, observe a red "blood spot." Note the small white spot (the position of the embryo) within the "blood spot." This area where the embryo develops is called the *blastodisc*. Examine the blastodisc with the dissecting microscope. As cell division occurs, a plate of cells will spread out from this area. The membrane that surrounds the yolk and the developing embryo is the *vitelline membrane*.

Surrounding the yolk is the "white" of the egg. It is composed of albumen, a protein material. Notice that the dense albumen forms an opaque cord on each end of the yolk. This is the *chalaza*. Locate the *air space* on the blunt end of the egg. At the air space, you can identify the *shell membrane* and the *inner egg membrane*. On the Data Sheet drawing, label the following structures: *shell, shell membrane, air space, chalaza, albumen, vitelline membrane, yolk, blastodisc*. Answer questions 1–5.

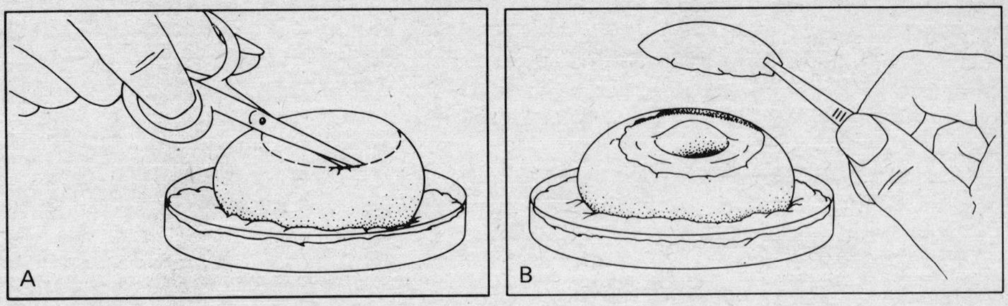

(B) Place a 60-hour incubated chick egg on a cotton bed in a fingerbowl. As before, cut and remove a lid so you can see inside. Remove some of the fluid albumen with an eye dropper so the embryo can be clearly seen. Use the dissecting microscope to examine the embryo. Label the following structures on the Data Sheet drawing: the developing brain, the neural tube (the developing spinal cord), somites (the paired sections of developing muscle tissue), the heart, and blood vessels. Answer questions 6–9.

(C) Place a 120-hour incubated chick egg on a cotton bed. Open it carefully and draw out some of the albumen with the eye dropper. Note the clear membrane, the *amnion*, that covers the embryo and forms a fluid sac called the *amniotic sac*. Another membrane grows from the embryo and surrounds the yolk — the *yolk sac*. Also find the small sac at the hind end of the embryo. This is the *allantois*. As development progresses, this will spread until it lines the entire shell.

Carefully lift the amnion with the forceps and cut it with the scissors to free the embryo. Examine the embryo with the dissecting microscope. Label the following structures on the Data Sheet drawing: brain, eye, limb buds, amnion, yolk sac, and allantois. Answer questions 10–13.

Investigations On Your Own

Culture frog eggs in the laboratory. Since the process of cell division occurs within a few hours after fertilization, various stages of development may be observed. Select eggs at an early stage of development and transfer them to a fingerbowl half filled with pond water and containing two sprigs of elodea. Place the fingerbowl in an area where the temperature will remain between 18°C and 20°C. Remove several eggs each day and examine them with the dissection and compound microscope. Make labeled diagrams to show the changes you observe in the development of tadpoles.

Discard any eggs that become cloudy and do not develop. Hatching tadpoles may be transferred to a large bowl and fed with chopped baby-food spinach as you observe their development. After your study, return the young animals to the pond where you collected the eggs.

NAME DATE

Chick Embryo Development 47

DATA SHEET

Analysis

In the space below, draw the parts of an unfertilized egg. Label the *yolk, blastodisc, albumen,* and *chalaza.*

Investigation

(A) Follow the instructions on your Procedure page and then label the drawing below. Identify the *shell, shell membrane, air space, chalaza, albumen, vitelline membrane, yolk*, and *blastodisc*.

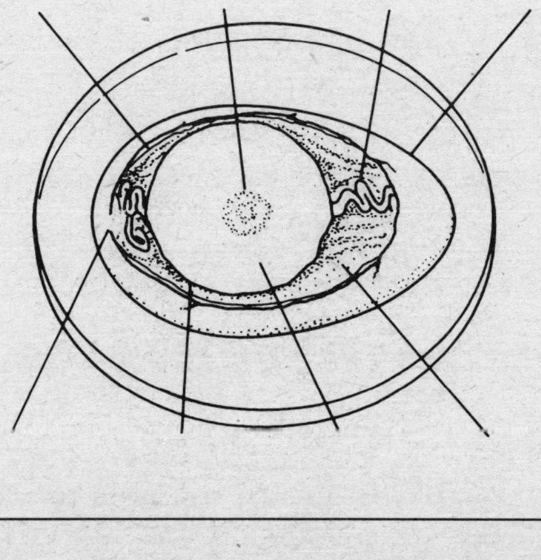

(1) What is the function of the albumen? ______________________________

____

(2) Where does the embryo develop? ______________________________

____

(3) What does the yolk provide for the embryo? ______________________________

(4) What structure holds the yolk in place? ______________________________

(5) What is the purpose of the shell and shell membranes? ______________________________

____

(B) Now examine a 60-hour incubated chicken egg as described on your Procedure page.

(6) Compare the position of the embryo with the position of the blastodisc you observed in **(A)**. ______________________________

____

Identify the developing *brain*, the *neural tube, somites, heart*, and *blood vessels* in the figure below.

60 Hour Chick Embryo

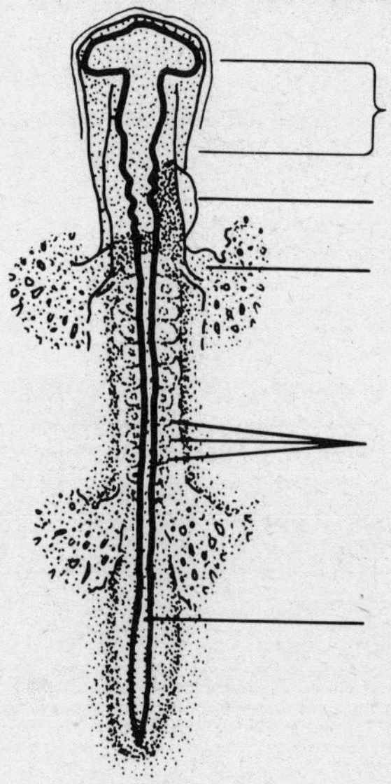

(7) Do you see any evidence of blood in the embryo? ______________________________

(8) If you do, describe the direction of flow. ______________________________

____

(9) What other primitive structures are you able to identify? ______________________________

____

NAME DATE

(C) Carefully examine a 120-hour incubated chicken egg as described on your Procedure page.

(10) What is the function of the amniotic sac? ______

(11) What structures do you see that identify the important function of this membrane forming the yolk sac? ______

(12) What is the function of the allantois? ______

(13) What features are becoming more defined? ______

In the drawing below, label the *brain, eye, limb buds, amnion, yolk sac*, and *allantois*.

120 Hour Chick Embryo

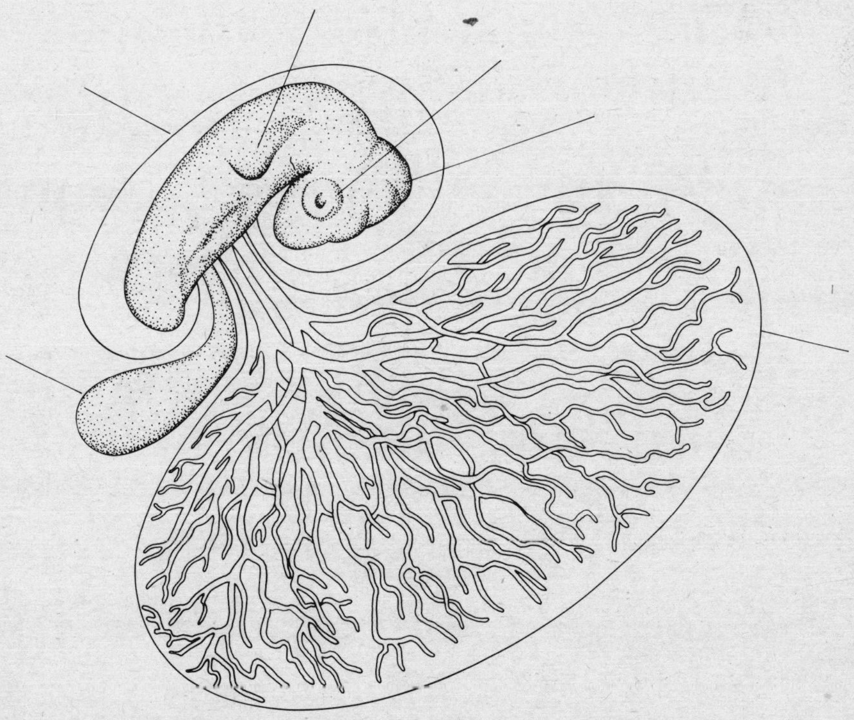

Summary

(1) What is the main difference you observed between the unfertilized egg and the fertilized, unincubated egg? ______________________

____

____

(2) How does a 120-hour incubated chicken embryo differ from a 60-hour incubated chicken embryo? ______________________

____

____

____

Test what you have learned by matching the term in the left column with the statement to which it is best suited. You may use a term more than once, and some you may not use at all.

a. air space
b. allantois
c. amnion
d. amniotic fluid
e. blastodisc
f. chalaza
g. chorion
h. neural tube
i. shell
j. somites
k. yolk sac

_______ **(3)** grows from digestive tract and surrounds yolk

_______ **(4)** form the muscles

_______ **(5)** secreted by membrane and encloses embryo

_______ **(6)** grows from digestive tract of embryo and functions in respiration and excretion

_______ **(7)** holds yolk and developing embryo in center of egg

_______ **(8)** prevents water loss but allows gas exchange

_______ **(9)** contains network of blood vessels and absorbs yolk

_______ **(10)** forms the spinal column and brain

_______ **(11)** rapidly dividing cells forming a plate at the surface of the yolk

NAME DATE

Interactions in the Biosphere 48

MATERIALS

ANALYSIS▶
- colored pencils

INVESTIGATION▶
- compound microscope
- glass slide and cover slip
- dropping pipet (30 centimeters long)
- Celsius thermometer
- pHydrion paper
- light meter (for incident light)
- covered, balanced aquarium containing water, gravel on bottom covered by finer sand, plants (*Elodea*, *Vallisneria*, and *Myriophyllum*), animals (2 or 3 snails, 1 small clam, 2 aquatic newts or 2 small goldfish)

OBJECTIVES▶
- To illustrate the water cycle by means of a diagram
- To understand the energy driving the water cycle
- To understand that cycles are involved in a stable ecosystem
- To discover some interrelationships among biotic and physical factors in an ecosystem

Problem

What is the water cycle and how does it work? Do interactions between and among biotic and physical factors occur in an ecosystem?

Water is used over and over again in a cycle. A water molecule falling from a rain cloud may enter a lake or an ocean. It may fall on the earth and be taken up by the root of a plant or drunk by an animal. But if we could follow a specific water molecule, we would see that it would eventually end up back in another rain cloud. This series of repeated events is called a *cycle*.

Several materials cycle in the biosphere. Nitrogen, carbon, oxygen, and water are important chemicals that cycle and involve living things. The cycles involve interactions among the physical factors of the biosphere, between the biotic and physical factors, and among biotic factors.

Procedure

Part I. Analysis

Turn to the Data Sheet and observe the drawing of a landscape. Think about the many things that can happen to a drop of water and indicate the water cycle, with arrows to show direction. When you have finished, refer to your text to see if you thought of all the possibilities. Then answer questions 1–6 on the Data Sheet.

Part II. Investigation

(A) Components of an aquarium as a small stable ecosystem

Examine the aquarium and identify its components. Think about which factors are biotic and which are physical. Observe that the gravel layer on the bottom appears to be darker in color than the sandy layer on top. It was not like this when the aquarium was first set up. Answer questions 7–10 on the Data Sheet.

Interactions between and among the biotic and abiotic factors also cause differences in the column of water above the gravel and sand. Use the Celsius thermometer and, using care not to stir up the water, observe the temperature of the water just below the surface, near the middle of the water column, and at the bottom. Record your findings in the chart on the Data Sheet.

With the dropping pipet, take a sample of water from just below the surface. Put one drop of water on the pHydrion paper to measure the pH and with another drop of the sample prepare a wet mount for microscopic examination. Using your text, identify the organisms present in the water. Record the pH and organisms in the chart on the Data Sheet.

Use the 30-centimeter pipet to remove a water sample from near the middle of the water column. Be careful not to stir up the water. Use pHydrion paper to measure the pH and again prepare a wet mount for microscopic examination. Record your findings in the chart on the Data Sheet.

Rinse out the 30-centimeter pipet. Then remove a water sample from very close to the bottom, using care not to stir up the water or remove any sand. Repeat the pH measurement and the microscopic examination of the sample. Record your findings in the chart on the Data Sheet.

Rinse out the 30-centimeter pipet. Remove a water sample from just below the surface of the bottom, using care not to stir up the bottom. Repeat the pH measurement and the microscopic examination of the sample. Record your findings in the chart on the Data Sheet.

(B) Energy entering the small, stable ecosystem
Point a light meter toward a window and notice the point on the scale where the indicator stops. Now, point the light meter towards the inside of the room and notice the point on the scale where the indicator stops. Answer question 11 on the Data Sheet.

Hold the light meter against the side of the aquarium that faces a window. Point the dial *toward* the window and measure the light intensity. Measure the light intensity of the other three sides of the aquarium. Answer questions 12–13 on the Data Sheet.

Investigations On Your Own

Determine the effect of relative humidity and temperature on the rate of evaporation. Select an outdoor station where a pan of water will not be disturbed for 24 hours. The station should be away from buildings or other obstructions that would interfere with air movement. Measure exactly 500 milliliters of water in a graduated cylinder and pour it into a shallow pan. (In hot weather, it may be necessary to use 1,000 milliliters of water.) After 24 hours, measure the water remaining in the pan, using the graduated cylinder. Record this amount in a table and determine the amount of water evaporated. Consult your newspaper or your local weather bureau to find the high and low relative humidity, high and low temperature, and high and low wind velocity during the 24-hour period. Record these data in your table. Repeat the procedure on 5 consecutive days, if possible. Write a report on your observations and conclusions.

NAME DATE

Interactions in the Biosphere 48

DATA SHEET

Analysis

Use the landscape shown below as a setting for making a diagram of the water cycle. You can indicate the movement of water, using arrows to show direction. Use *red* to indicate precipitation, *blue* for surface water, *green* for ground water, and *yellow* for water evaporating and returning to the atmosphere. Indicate, on the arrows, the process that is occurring in each step of the water cycle you drew.

(1) What is the process of water vapor formation called? ______________________

(2) What is the process by which water returns to the earth? ______________________

____

(3) What can happen to a raindrop that falls to the earth? ______

______

______

______

(4) Explain the "driving force" of the water cycle in the biosphere. ______

______

______

______

______

______

(5) Explain the role of plants in the water cycle as an interaction between biotic and physical factors. ______

______

______

(6) Explain, as an interaction between physical factors, how runoff water may cause erosion to topsoil and alter the normal water cycle. ______

______

______

______

Investigation

(A) Components of an aquarium as a small, stable ecosystem

(7) Name the plants you observe growing in the aquarium. ______

______

(8) Name the animals you observe in the aquarium. ______

______

(9) Name the physical components of the aquarium. ______

(10) Can you explain the difference in the gravel color at the bottom of the aquarium?

______

______

NAME DATE

Water column	Temperature	pH	Organisms
Near surface	1.		
	2.		
	3.		
	4.		
Middle	1.		
	2.		
	3.		
	4.		
Near bottom	1.		
	2.		
	3.		
	4.		
In bottom material	1.		
	2.		
	3.		
	4.		

(B) Energy entering the small, stable ecosystem

(11) How does the light meter indicate light intensity by the value it shows? ______________

____

____

(12) How does the light intensity differ on the three sides of the aquarium? ______________

____

(13) What significance does light intensity have for the plants and animals in the aquarium? ______________

____

____

____

Summary

(1) How do you think the physical factors of the small, stable ecosystem differ from those of the newly set-up aquarium? ______

(2) What interactions between the biotic and physical factors were involved in your observations? ______

(4) What interactions between the biotic factors were involved in your observations? ______

(4) Describe two cycles that were operating in the small, stable ecosystem of the aquarium. ______

NAME DATE

Ecosystems and Energy 49

MATERIALS

INVESTIGATION▶

- covered balanced aquarium containing water, coarse sand on bottom covered by finer sand, plants (*Elodea, Vallisneria,* and *Myriophyllum*), animals (2 or 3 snails, 1 small clam, 2 aquatic newts or 2 small goldfish)

OBJECTIVES▶

- To view an ecosystem as a unit of the biosphere
- To identify the relationships of the components of an ecosystem
- To understand productivity
- To understand energy transfer in trophic levels

Problem

What is an ecosystem, and how is energy involved?

An ecosystem is a complex unit of the biosphere in which living and nonliving things interact. Biotic communities and the physical environment are characteristics that differ in various ecosystems. The boundaries of an ecosystem are not always distinct as in an aquarium. However, materials do cycle within the aquarium, and energy relationships can be examined.

All ecosystems in the biosphere receive energy from the sun. The organization of an ecosystem involves *producers* that trap the sun's energy and store it as chemical energy. The producers form the base of a *food chain* through which energy is passed on from one organism to another. But most energy consumed is *not* stored. It is used up in supporting life activities.

Food chains are not simple, because most animals eat more than one type of food. The interrelations of food chains form *food webs*.

Procedure

Part I. Analysis

Begin your analysis of an ecosystem by looking at the diagram of a small, stable ecosystem at the top of the Data Sheet. Compare this diagram to your small, stable ecosystem in the aquarium. Answer questions 1–3 on the Data Sheet.

Part II. Investigation

Look at the balanced aquarium in your classroom. Think about the energy transfer relationships that are taking place in this small ecosystem. Compare the organisms you have identified to those shown in the Analysis diagram at the top of the Data Sheet. Answer questions 4–6 on the Data Sheet.

The analysis of energy transfer in an ecosystem is often represented as a *food pyramid* like that shown in the diagram on the next page. An actual food chain in your aquarium is indicated by the organisms in the right section of the pyramid. Consider this example, with the food chain beginning with the unicellular algae, and answer question 7 on the Data Sheet.

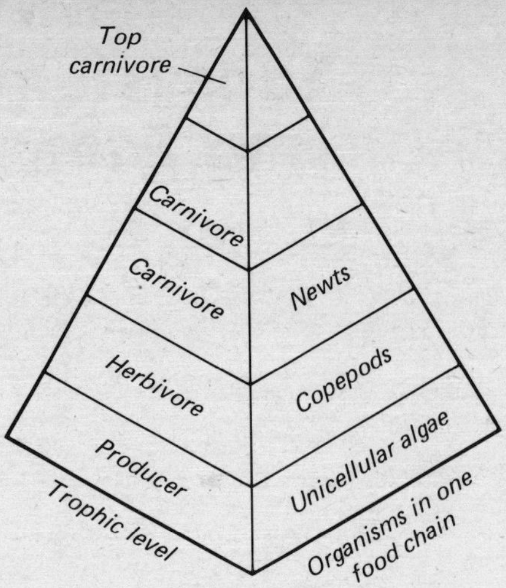

At any given time, the organisms forming the base of the food pyramid represent the energy available to the system. When herbivores consume plants, they accumulate about 15 percent of the energy taken in. In turn, carnivores accumulate about 15 percent of the energy they consume from the herbivores. Consider one food chain — unicellular algae–copepods–newts — in your aquarium. Answer questions 8–12 on the Data Sheet.

The energy transfer process can be applied to any one food chain, or to the total *biomass* of any of the trophic levels in an ecosystem (such as your aquarium). The accumulation of energy in an ecosystem is called *primary productivity*. Think about energy transfer in your aquarium and answer questions 13–15 on the Data Sheet.

Scavengers were not indicated in the food pyramid. They do, however, play an important role in an ecosystem. Answer questions 16–17 on the Data Sheet.

Decomposers were also not indicated in the food pyramid. Think about their role in the aquarium and answer question 18 on the Data Sheet.

Investigations On Your Own

(1) Choose another ecosystem, such as coniferous forest, deep ocean, or tundra, and construct the food chains and food webs that exist in that system.

(2) Select two totally different environments near your home. Record variations in temperature, light, and air movement. Examine and classify the soil and test its pH. Collect specimens of plants and observe the various animals in the habitat. List the organisms as autotrophs, heterotrophs, and carnivores. Construct food webs to compare these two habitats.

NAME DATE

Ecosystems and Energy 49

DATA SHEET

Analysis

Compare this diagram to the small, covered aquarium you have studied in the classroom. Then answer questions 1–3.

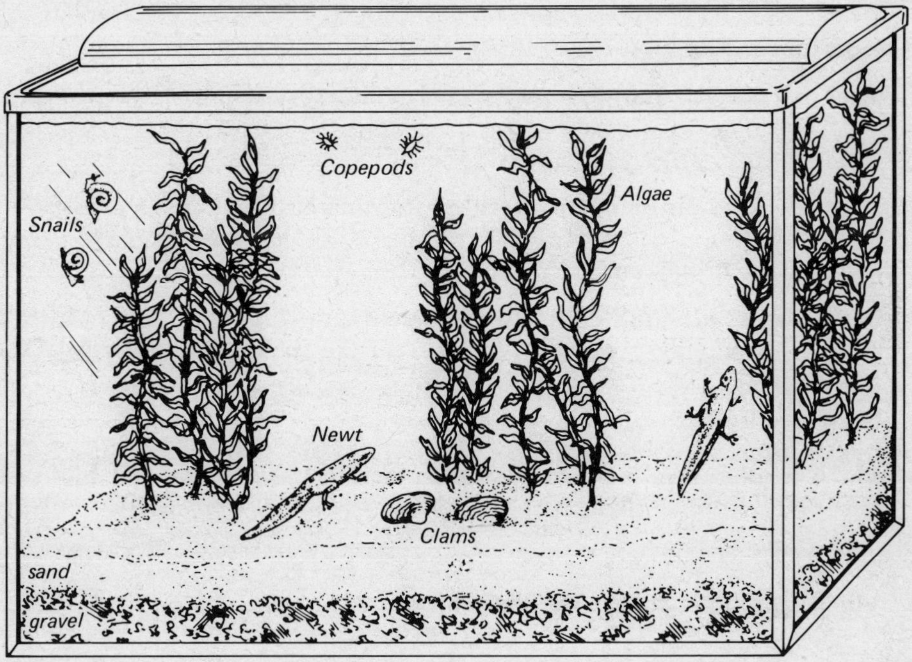

(1) Which organisms are food producers? ______

(2) Which organisms are herbivores? ______

(3) Which organisms are carnivores? ______

Investigation

(4) Identify several food chains taking place in the aquarium. ______

______

______

______

(5) What trophic levels are represented in the food chains you have identified? ______

______

(6) What trophic level is missing from the aquarium ecosystem? ______

______

(7) Why is the "pyramid of energy" a good term for the food pyramid illustrated?

____

____

(8) If a consumer stores only about 15 percent of the energy it takes in, what happens to the remaining 85 percent? ______________________________

____

____

(9) How much of the unicellular algae's energy is stored in the copepods of the aquarium in the food chain illustrated? ______________________________

(10) What percent of the copepod's energy is stored in the newts of the aquarium?

____

(11) In this food chain, what percent of the unicellular algae's energy is stored in the newts of the aquarium? ______________________________

(12) Another possible food chain in the aquarium might be: unicellular algae — paramecia — copepods — newts. What percent of the original energy of the algae would be stored in the newts? ______________________________

(13) Which trophic level in your stable aquarium has the greatest biomass? __________

____

(14) Which trophic level has the least biomass? ______________________________

(15) How could you measure the carnivore biomass in your aquarium? __________

____

(16) How were scavengers identified in the small, stable ecosystem of the aquarium?

____

(17) Which scavengers are also herbivores? ______________________________

(18) Name the decomposers in the aquarium and explain their role. __________

____

Summary

(1) A food chain involves energy transfer. Does energy cycle? Explain. __________

____

____

(2) In an ecosystem, where is the greatest amount of energy stored? __________

____

NAME DATE

Changing Biotic Communities 50

MATERIALS

INVESTIGATION▶

- pond water
- timothy hay
- wheat grains
- refrigerator
- 2 1-liter bottles
- 5 250-milliliter beakers
- 3 Petri dishes
- wax marking pencil
- graduated cylinder
- Celsius thermometer
- compound microscope
- microscope slide and cover slip
- pHydrion paper
- eye dropper
- colored pencils

OBJECTIVES▶

- To become aware of many niches and interactions in a habitat
- To observe biotic changes during succession in a small ecosystem
- To observe some physical changes occurring in a small ecosystem
- To see how population densities change during succession

Problem

Can community changes be observed during succession?

The organisms naturally occurring within an ecosystem react with one another and with their environment. If a new ecosystem could be set up, certain organisms would soon become very numerous. These are the *pioneers*. They change the environment and make it possible for other organisms to live in the ecosystem. This variety of organisms makes up the *biotic community*.

To survive, organisms must be able to adjust to these changes. Patterns within communities can be observed. The changing of communities is called *succession*.

Procedure

Part I. Analysis

In any ecosystem, the organisms interact with one another and with the physical environment. The *habitat* of an organism is the place where it lives. The *ecological niche* is the way in which it "earns a living" or obtains its energy. Turn to the Data Sheet and look at the diagram showing several organisms living in a prairie. Think about the various niches and answer questions 1–8 on the Data Sheet.

Part II. Investigation

Interactions of organisms alter the environment. When organisms invade a new area, their effect can be observed. Changes occur in the numbers and kinds of organisms present. You can examine some of these *succession* changes in pond water.

Set up small cultures called *hay infusions*. These cultures are small ecosystems in which you can observe community changes that occur during succession.

Boil 2 liters of pond water. Allow it to cool, and transfer it to two 1-liter bottles. These can be stored in the refrigerator until required. Number the five 250-milliliter beakers 3, 6, 9, 12, and 15. Now measure out 200 milliliters of the boiled pond water and put it into beaker 15. Add 20 pieces of timothy hay (each 3 centimeters long) and 4 wheat grains. Stir the mixture and examine a drop of it with the compound microscope. Answer question 9 on the Data Sheet. Then record your findings for "0 days" in the table on the Data Sheet. Describe the appearance, pH, and temperature of the beginning culture. Cover the beaker with half of a Petri dish, and set the beaker in a well-lighted place in the room.

Three days later, repeat this procedure with beaker 12. You may omit the microscopic examination of the pond water. Continue to make a new culture every 3 days. One or two beginning culture days will fall on a weekend, depending on the day this Investigation is started. You may start these cultures on the day before or the day after the weekend. On the 12th day, begin the hay infusion in beaker 3. On the 15th day, you will be ready to make your observations.

Observe the appearance and note any odor in each of the cultures. Using pHydrion paper, measure the pH of each culture. Measure the temperature of each culture. Record these findings in the table on the Data Sheet.

Now, stir up the contents of the 3-day-old culture. Remove a drop (avoid any solid material) and prepare a slide for microscopic examination. Remember that various bacteria will probably be growing in your culture now, but it's unlikely you will observe them, because of their small size. Answer questions 10–12 on the Data Sheet.

As you search the slide under the microscope, use a definite pattern to avoid counting the same organism twice. Keep a tally of the organisms you identify according to their specific groups. Groups easily identified are: green protists (Producers), nongreen protists, rotifers, crustaceans, annelids, and roundworms. If you observe a specific organism, such as *Euglena* or *Paramecium*, you can add it to your list of protists. When you have covered your slide once, record your observations in the chart on the Data Sheet.

Clean your glass slide, cover glass, and eye dropper well before going on to the "day 6" culture. Repeat the inspection and the microscopic survey of this culture. Record your observations in the table on the Data Sheet. Then repeat this procedure with the remaining cultures.

When you have completed your inspection of the cultures, take a class average count to increase your sample size. Use the class average counts per drop as an *index of population*. To show the observed changes in community structure of your small ecosystem, plot these data to make a graph on the Data Sheet. You may use colored pencils to represent the various individuals. *Be sure to identify the color with the organism*. Use ink to plot the combined class average number of all organisms present. Answer questions 13–17 on the Data Sheet.

Investigations On Your Own

(1) Repeat this Investigation but maintain the cultures in the dark. Make a graph of your results.
(2) Repeat this Investigation but maintain the cultures at two different temperatures. Compare the results and make graphs to show how the cultures differ.

NAME DATE

Changing Biotic Communities 50

DATA SHEET

Analysis

Look at the diagram below, which shows many of the organisms living on the prairie. Consider the various ecological niches and answer questions.

(1) Which organisms are autotrophs? ______

(2) Which organisms are heterotrophs? ______

______

______

(3) Which organisms are herbivores? ______

______

(4) Which organisms are carnivores? ______

______

(5) Which organisms are top carnivores? ______

(6) What name is given to all the organisms in the prairie ecosystem? ______

______

(7) What is the niche of the coyote? ______

______

(8) Could there be any overlapping of the niches of the burrowing owl and the hawk?

______

______

Investigation

(9) Describe the appearance, pH, temperature, and any organisms in the drop of the hay infusion when it was first made. ______

______

(10) Why is it important to consider the presence of bacteria in the hay infusion?

______

(11) How did the bacteria get into the culture? ______

______

(12) Where do the bacteria get the energy they need to grow? ______

______

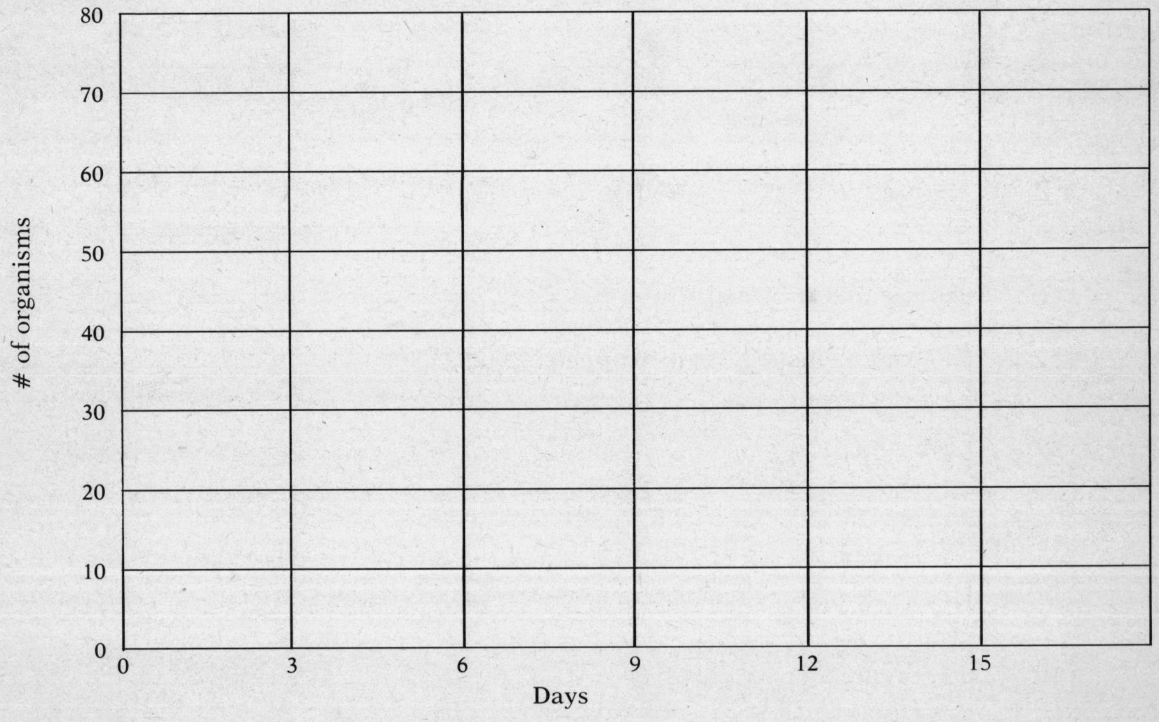

(13) What were the most obvious changes you observed in the kinds of organisms present at different ages of the cultures? ______

______

(14) What were the pioneer organisms? ______

NAME DATE

(15) How did the organisms that followed the pioneers get into the culture? ____________

____

HAY INFUSION OBSERVATIONS					
Age (days)	Appearance	pH	Temp. °C	Number of Organisms per Drop	Class Average of Organisms per Drop
0					
3					
6					
9					
12					
15					

(16) What organisms comprise the climax community in your culture? ____________

____

(17) Were any of these organisms observed in the 3-day-old culture? in the 15-day-old culture? ____________

Summary

Write a paragraph to summarize your study of succession in the small ecosystem formed by the hay infusion. Think about such terms as: pioneer, dominant, succession, ultimate energy source, autotrophs, heterotrophs, transfer of energy, food chain, pond water, timothy hay, and wheat grains.

NAME DATE

Population Growth 51

MATERIALS

INVESTIGATION▶

- molasses
- dried yeast
- graduated cylinder
- 1-liter flask
- eye dropper
- cell counting chamber with cover slip
- microscope
- incubator set at 25° Celsius (optional)

OBJECTIVES▶

- To measure a population by sampling
- To determine a population growth curve
- To understand some factors limiting the growth potential of a population

Problem

How can a population be measured and what factors affect its size?

The biotic community of an ecosystem is made up of many populations. A *population* is a group of individuals of one species. Population measurements may be made by taking samples of an area and actually counting the individuals of a species in the sample. The number counted per unit area is the *density*. The density is then used as an index to the size of a population in an ecosystem.

It is possible to measure population growth in a yeast culture. A known number of individuals can be introduced into the culture. Then, by taking samples, the individuals making up the whole population can be estimated. As the yeast cells grow, you can see some ways in which the environment changes.

Procedure

Part I. Analysis

Turn to the Data Sheet. Examine the first population growth curve, which shows growth of a sheep population. Describe what the curve shows by answering questions 1–3.

Examine the second population growth curve, which shows growth of a plankton population over a period of one year. Describe what the curve shows by answering questions 4–6.

Part II. Investigation

(A) Sampling a yeast population

In a 1-liter flask, mix together 250 milliliters of light molasses with 250 milliliters of water. Add about ¼ package of dried yeast and mix well. Place the culture in a warm place (not in direct sunlight) or in an incubator set at 25° Celsius. After 30 minutes, mix the contents well by swirling the flask and then remove a 1-drop sample. Put the sample on the ruled slide and add a cover slip. Using the ruled counting slide, count the number of yeast cells in 0.1 cubic millimeter (0.1 milliliter) and figure the density of yeast cells (number per milliliter). Then compute the *beginning* population in the yeast culture. Use individual data to compute an average for the class. (See the example on the following page.) Record your conclusions in the chart on the Data Sheet.

Repeat your measurements each day for 15 days. Record your conclusions in the chart on the Data Sheet. Don't forget to swirl the culture, thoroughly mixing the contents *before* removing your sample. Since you will not make counts during weekends, some data will be missing.

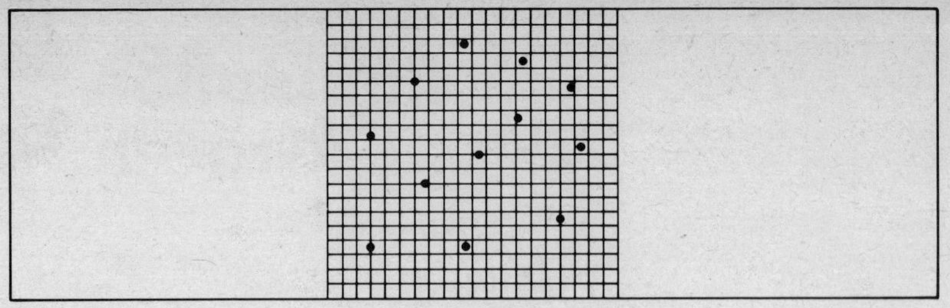

ruled counting slide showing 12 yeast cells in 0.1 milliliters

(0.1 ml x 12 = 1.2/ml x 250 ml = 300 yeast cells in beginning population)

(B) Graphing results of a growing population of yeast cells
On the Data Sheet, label the time (days) on the horizontal axis (ordinate). Label the number of cells (population) on the vertical axis (abcissa). You may wish to label your population in "thousands of cells." In pencil, plot your individual data. Use a pen to plot the average class data.

(C) Conclusions from a study of a growing population of yeast cells
Carefully examine the results of your yeast population growth study in your graph. Then, answer questions 7–9 on the Data Sheet.

Investigations On Your Own

(1) Use the sampling technique to estimate the population of individual grass plants in the school football field.
(2) Take samples of the same size from various places in an empty lot. One way to do this is to mark off a plot 30 centimeters square and remove the soil to a depth of 5 centimeters. With a hand lens and dissecting and compound microscopes, examine the soil carefully. Identify as many organisms as you can and keep a record of their numbers. Record the type of soil, volume of sample, number of organisms, and number of organisms per cubic centimeter.
(3) The suggested research above is interesting as a team research project. Various areas of the same plot can be examined by different teams and the results compared. Perhaps you can identify food chains that operate in the area. Another suggestion would be for different teams to make such measurements at various times of the year. Comparative results may show that some populations vary seasonally.

NAME ____________________ DATE ____________________

51

Population Growth

DATA SHEET

Analysis

Growth of a sheep population on an island that had no previous sheep population:

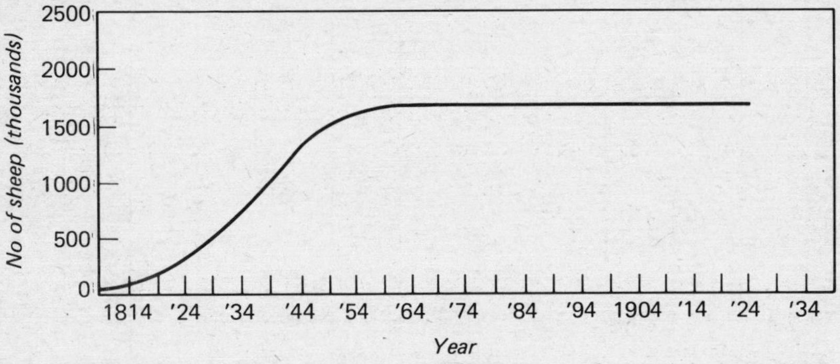

(1) Describe the growth of the sheep population. ____________________

(2) What probably happened to the sheep population in 1934? ____________________

(3) What factors might have affected the growth of the sheep population? ____________________

Plankton growth curve over a period of one year:

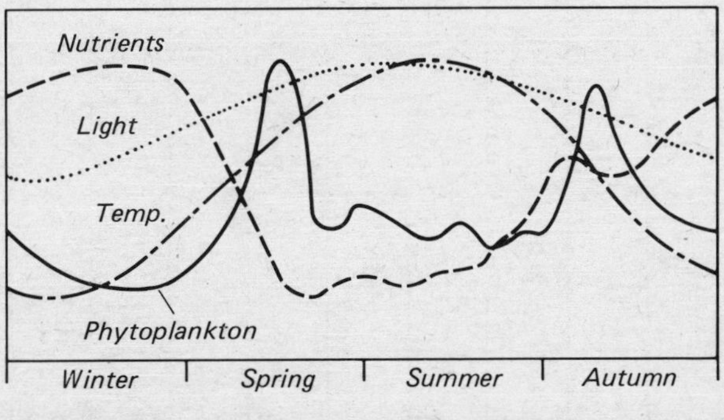

(4) Describe the growth of a plankton population over a year. ____________________

(5) What factors are shown that influence the growth pattern of plankton? ____________________

(6) What might be a reason for the decrease in the population in the summer? ____________________

Investigation

(A) Sampling a yeast population

RECORD OF YEAST POPULATIONS

days

	1	2	3	4	5	6	7	8	9	10	11	12	13	14	15
Yeast cells counted															
Density/ mL															
Population of yeast cells															
Class average of population															

(B) Graphing results of a growing population of yeast cells

NAME DATE

(C) Conclusions from a study of a growing population of yeast cells

(7) Explain the results indicated by the curve on the graph for the first 4 or 5 days.

(8) Explain the results indicated by the curve on the graph from day 5 to day 10.

(9) Explain the results indicated by the curve on the graph from day 11 to day 15.

Summary

(1) Did the yeast cells you investigated reach their reproductive potential? Explain your answer.

(2) Explain why a "carrying capacity" was not involved in your Investigation of the yeast population.

NAME DATE

52 Natural Resources, Energy, and Pollution

MATERIALS

INVESTIGATION▶

- pond water
- timothy hay
- wheat grains
- refrigerator
- 10 1-liter bottles
- 25 250-milliliter beakers
- 13 Petri dishes
- wax marking pencil
- motor oil
- glass stirring rod
- graduated cylinder
- Celsius thermometer
- compound microscope
- microscope slide and coverslip
- pHydrion paper
- eye dropper
- colored pencils

OBJECTIVES▶

- To realize that human activities alter the environment
- To understand that some natural resources are renewable and some are not
- To realize that some energy sources are limited
- To observe effects of oil pollution on organisms in an environment
- To examine effects of oil pollution on succession

Problem

What are some environmental changes brought about by human activity? Interactions between organisms and their environment bring about changes in the environment. Food chains interact in an ecosystem and can be studied as food webs. Human activities cause changes that interfere with food chains by adding pollution and by depleting available energy sources of the earth.

Although every organism brings about changes in the environment, humans have the ability to analyze these changes. Humans can control the extent to which they are willing to allow the environment to be altered. Humans also recognize that changes will affect *all* organisms sharing the environment.

Human activities often cause unnecessary pollution. One common pollutant is oil. When oil enters the water, it does not mix readily. It sticks to objects and, as it breaks down, some of its components are dissolved by water. Oil is spilled by boats and ships. Oil dripping from automobiles onto the roadways is washed into storm sewers by rain and enters streams, rivers, lakes, and oceans. If you change the motor oil in your car and pour the used oil into the gutter, think about where it eventually ends up!

Procedure

Part I. Analysis

Look at the diagram at the top of the Data Sheet. Notice the interaction of energy, a natural cycle, and human activities. Fill in the chart below the diagram. Answer questions 1–3.

Part II. Investigation

Set up five sets of hay infusions as in Investigation 50. Begin by numbering five 250-milliliter beakers A–3, A–6, A–9, A–12, and A–15. These will serve as your *control*. You might expect the same results as in Investigation 50, but a control must be made with every experiment. The *experimental variable* is motor oil that will serve as a *pollutant*. Since you do not know how much oil to add to each culture, you will test *four* different concentrations and establish a *dose curve* to determine pollution effects of motor oil on succession.

Number five 250-milliliter beakers B–3, B–6, B–9, B–12, and B–15.

Number five 250-milliliter beakers C–3, C–6, C–9, C–12, and C–15.

Number five 250-milliliter beakers D–3, D–6, D–9, D–12, and D–15.

Number five 250-milliliter beakers E–3, E–6, E–9, E–12, and E–15.

Add 20 pieces of timothy hay (3 centimeters each) and 4 wheat grains to the beakers labeled A–15, B–15, C–15, D–15, and E–15.

Add 0.5 milliliters of motor oil to beaker B–15.

Add 1.0 milliliters of motor oil to beaker C–15.

Add 2.5 milliliters of motor oil to beaker D–15.

Add 3.0 milliliters of motor oil to beaker E–15.

Swirl the beakers containing motor oil, timothy hay, and wheat grains to thoroughly mix the contents. Next, add 200 milliliters of boiled, cooled, pond water to beakers A–15, B–15, C–15, D–15, and E–15. Answer questions 5 and 6 on the Data Sheet. Then record your findings for "0 days" in the Table on the Data Sheet. Describe the appearance, pH, and temperature of the beginning culture. Cover each beaker with half of a Petri dish and set the beakers in a well-lighted place in the room.

Three days later, repeat the above procedure with beakers A–12, B–12, C–12, D–12, and E–12. Continue to make new cultures every 3 days. If a beginning culture day falls on a weekend, you may start the culture on the day before or the day after the weekend. On the 12th day, you will begin the hay infusions in beakers A–3, B–3, C–3, D–3, and E–3. On the 15th day, you will be ready to make your observations.

Observe and compare the appearance and note any odor in each of the cultures. Using pHydrion paper, measure the pH of each culture. Measure the temperature of each culture. Record these findings in the Table on the Data Sheet.

Using a clean stirring rod, mix the contents of each 3-day-old culture. Prepare slides for microscopic examination and compare the cultures A, B, C, D, and E. Keep a tally of the organisms you identify according to their specific groups as you did in Investigation 50 but, on the graph on the Data Sheet, record the total number of all organisms present. Use a black pencil for set A, blue for set B, green for set C, orange for set D, and red for set E.

Clean your glass slides, cover glasses, and eye droppers well before going on to the "day 6" culture. Repeat the inspection and the microscopic survey of this culture. Record your observations in the Table on the Data Sheet. Then repeat this procedure with the remaining cultures.

When you have completed inspecting the cultures, take class average counts per drop. Plot these class averages on the graph on the Data Sheet by using the same colors as above, but use dotted lines. Answer questions 7–9 on the Data Sheet.

Investigations On Your Own

This Investigation may be repeated using a change in the concentration of motor oil in the cultures. It may also be repeated using a different pollutant. Detergents, insecticides, herbicides, and thermal pollution may be considered.

NAME _______________________________ DATE _______________

Natural Resources, Energy, and Pollution

52

DATA SHEET

Analysis

Study the diagram below, fill in the chart, and then answer the questions.

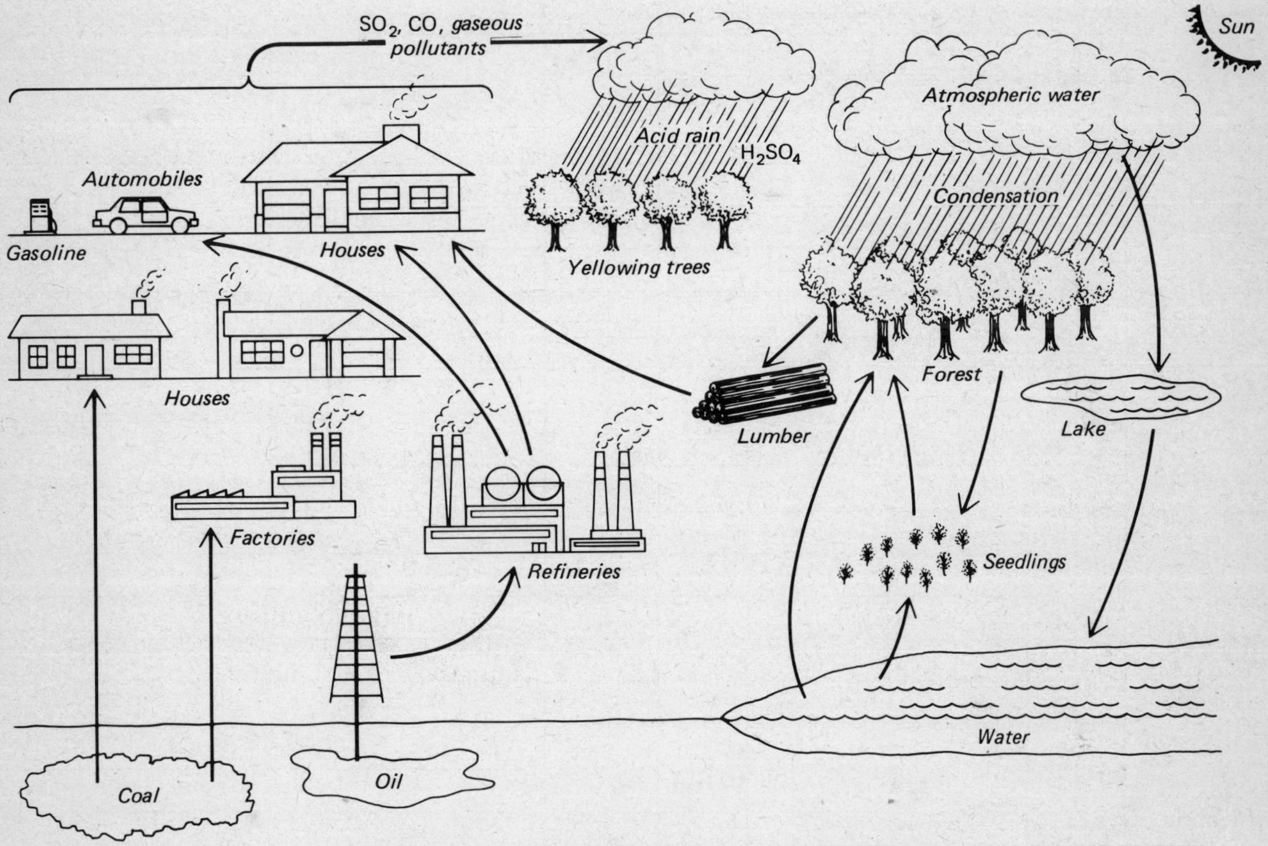

Cycles shown in diagram	Renewable resources in diagram	Nonrenewable resources in diagram	Human activity pollutants in diagram	Sources of pollution in diagram

(1) What are the Sources of sulfur dioxide (SO_2) and carbon monoxide (CO)? _______________

____

(2) How do acid rains form? ____

____

(3) How is the use of nonrenewable resources affecting the renewable resources? _______

____

HAY INFUSION OBSERVATIONS						
Age (days)		Appearance	pH	Temp. °C	Number of Organisms per Drop	Class Average of Organisms per Drop
0	A					
	B					
	C					
	D					
	E					
3	A					
	B					
	C					
	D					
	E					
6	A					
	B					
	C					
	D					
	E					
9	A					
	B					
	C					
	D					
	E					
12	A					
	B					
	C					
	D					
	E					
15	A					
	B					
	C					
	D					
	E					

NAME DATE

(4) If most of the spilled oil floated on the surface, how might it affect the fish that swim below? ______________________________

____

____

(5) Describe the appearance, pH, temperature, and any organisms in the drops of hay infusions in the cultures when first made. ______________________________

____

____

____

(6) State the "pollution concentration" of the motor oil in parts per thousand (ppt) and in parts per million (ppm) for each set of beakers. ______________________________

____

____

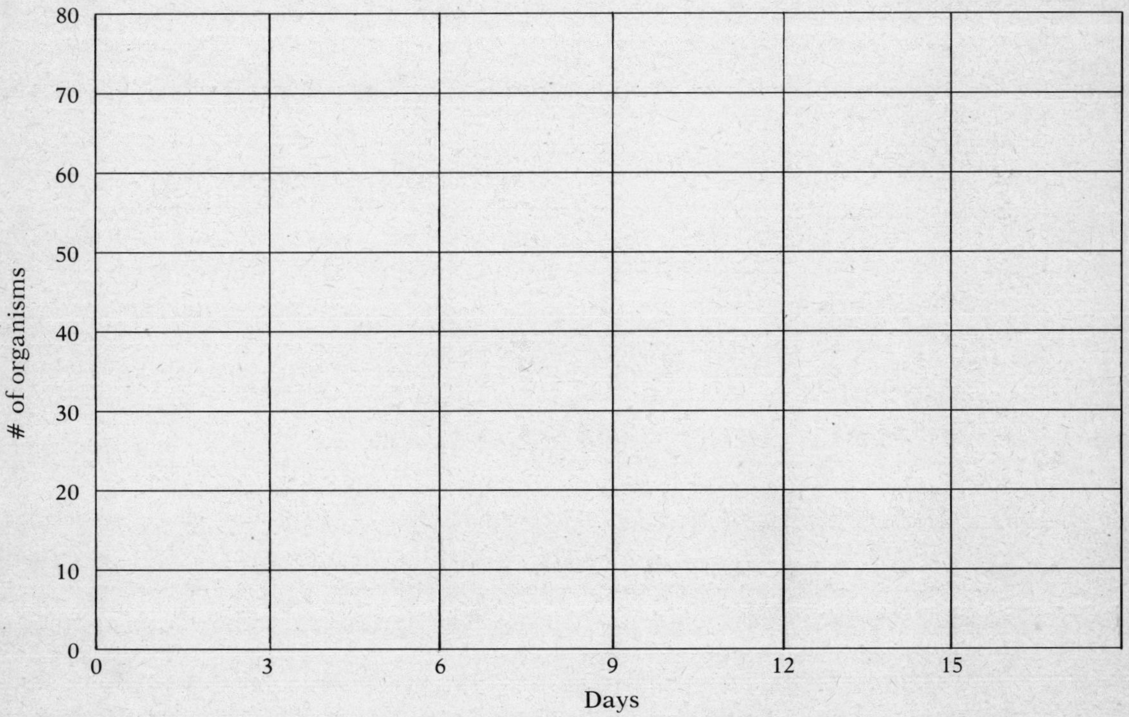

(7) How did the observations on the control of this experiment compare with the results of your baseline study in Investigation 50? ______________________________

____

(8) How did the total numbers of organisms differ in A, B, C, D, and E?

(9) Compare the observed succession of organisms in each of the experiments.

Summary

Write a paragraph to summarize this Investigation and draw conclusions regarding the effects of oil pollution from human activities.